Buetti-Dinh Antoine

Gene Regulation by Numbers

Buetti-Dinh Antoine

Gene Regulation by Numbers

Steady-State and Equilibrium Binding Applied to
Gene Regulation Systems

Südwestdeutscher Verlag für Hochschulschriften

Impressum/Imprint (nur für Deutschland/only for Germany)
Bibliografische Information der Deutschen Nationalbibliothek: Die Deutsche Nationalbibliothek verzeichnet diese Publikation in der Deutschen Nationalbibliografie; detaillierte bibliografische Daten sind im Internet über http://dnb.d-nb.de abrufbar.
Alle in diesem Buch genannten Marken und Produktnamen unterliegen warenzeichen-, marken- oder patentrechtlichem Schutz bzw. sind Warenzeichen oder eingetragene Warenzeichen der jeweiligen Inhaber. Die Wiedergabe von Marken, Produktnamen, Gebrauchsnamen, Handelsnamen, Warenbezeichnungen u.s.w. in diesem Werk berechtigt auch ohne besondere Kennzeichnung nicht zu der Annahme, dass solche Namen im Sinne der Warenzeichen- und Markenschutzgesetzgebung als frei zu betrachten wären und daher von jedermann benutzt werden dürften.

Coverbild: www.ingimage.com

Verlag: Südwestdeutscher Verlag für Hochschulschriften GmbH & Co. KG
Heinrich-Böcking-Str. 6-8, 66121 Saarbrücken, Deutschland
Telefon +49 681 37 20 271-1, Telefax +49 681 37 20 271-0
Email: info@svh-verlag.de

Approved by: Zurich, University of Zurich, PhD Dissertation, 2011

Herstellung in Deutschland (siehe letzte Seite)
ISBN: 978-3-8381-3288-4

Imprint (only for USA, GB)
Bibliographic information published by the Deutsche Nationalbibliothek: The Deutsche Nationalbibliothek lists this publication in the Deutsche Nationalbibliografie; detailed bibliographic data are available in the Internet at http://dnb.d-nb.de.
Any brand names and product names mentioned in this book are subject to trademark, brand or patent protection and are trademarks or registered trademarks of their respective holders. The use of brand names, product names, common names, trade names, product descriptions etc. even without a particular marking in this works is in no way to be construed to mean that such names may be regarded as unrestricted in respect of trademark and brand protection legislation and could thus be used by anyone.

Cover image: www.ingimage.com

Publisher: Südwestdeutscher Verlag für Hochschulschriften GmbH & Co. KG
Heinrich-Böcking-Str. 6-8, 66121 Saarbrücken, Germany
Phone +49 681 37 20 271-1, Fax +49 681 37 20 271-0
Email: info@svh-verlag.de

Printed in the U.S.A.
Printed in the U.K. by (see last page)
ISBN: 978-3-8381-3288-4

Copyright © 2012 by the author and Südwestdeutscher Verlag für Hochschulschriften GmbH & Co. KG and licensors
All rights reserved. Saarbrücken 2012

Steady-State and Equilibrium Binding Applied to Gene Regulation Systems

Dissertation
Zur
Erlangung der naturwissenschaftlichen Doktorwürde
(Dr. sc. nat.)

vorgelegt der
Mathematisch-naturwissenschaftlichen Fakultät
der
Universität Zürich

von
Antoine Buetti-Dinh
aus der
Schweiz

Promotionskomitee
Prof. Dr. Attila Becskei (Vorsitz, Leitung der Dissertation)
Prof. Dr. Nathan Luedtke

Zürich 2011

SUMMARY

This thesis is composed of three parts that are presented in a hierarchical order reflecting increasing organization level of the studied system. Gene regulation was studied at the single gene level first, extended to interacting transcriptional units and finally to its role within small regulatory networks.

The first part of this study has been centered on single transcription units of regulatory networks and the mechanisms underlying their control. Promoter sequences can in fact be regarded as control building blocks of genes that modulate transcription by integrating different cellular signals and their activity can be described mathematically using tools inspired by general concepts widely applied in different fields of science and engineering. Our approach consists of an iterative experimental-based mathematical modeling based on promoter libraries with synthetic binding site sequences of variable affinities for selected transcriptional activators. We used such libraries to perform *in vivo* experiments using the yeast *Saccharomyces cerevisiae* in order to improve our theoretical models. The main goal of this part of the work is the determination of gene system's intrinsic properties like cooperativity coefficients and binding affinities of transcription factors to DNA regulatory sequences. The technique developed for this part of the work provides a new way to study the regulation of genes *in vivo* over a broad range of activities with minimally invasive impact on cellular metabolism. We could retrieve the activation profile of *GAL1* promoter, a model system for gene regulation, over a broad range of activity. *GAL1* promoter activation profile was characterized by high reactivity at low induction but graded response at higher activity. Multiple binding sites contributed a more-than-additive increase of expression compared to the promoter activated by a single binding site revealing that activators can cooperate not by direct interaction but through components of the transcriptional machinery.

In the second part of the study we extended the approach to interacting transcriptional processes. Transcriptional interference is defined as the suppressive influence of one transcriptional process, directly and *in cis*, on a second transcriptional process. This mechanism of regulation involves antagonistic relationship between transcription units and has been shown to have implications in both yeast and higher eukaryotes. In transcriptional interference, the signals received by the two antagonistically acting activators are combined by the polymerase trafficking along the DNA. The principles of this signal conversion and how these signals can be utilized to control gene expression have been explored by designing a dual-control genetic system in yeast and subsequently interpreted using equilibrium and non-equilibrium models of transcription. Antagonism by an upstream activator displayed the characteristics of competitive inhibition, whereas a downstream activator inhibited gene expression in a non-competitive manner. Cooperative effects between transcriptional regulators could explain positive effects observed at low gene expression induction, revealing that antagonistic activators have the potential to drive paradoxical gene activation.

In the third part, the process of gene regulation has been further studied in the context of small regulatory networks. Particular attention was payed to feedback loops in the yeast regulatory network of the galactose metabolism. Based on the knowledge gained in the previous parts of the work, the properties of promoters have been further investigated to shed light on the role played by intracellular parameters within interconnected genetic circuits. The use of synthetic gene circuits introduced in yeast cells together with stochastic modeling allowed us to interpret more efficiently the role of parameters that influence the dynamics of gene expression. Different yeast promoters (*GAL1* and *GAL3*) that diverged from a common ancestral sequence could reveal mechanisms of evolutionary optimization to variable environmental conditions. The gene responsible for nutrient sensing displayed fast bistable state transition thereby improving adaptation to changing nutrient conditions. In contrast, the gene involved in the metabolic branch of the GAL network showed higher memory to previously applied conditions as a consequence of activator binding cooperativity and reduced basal expression.

ZUSAMMENFASSUNG

Die vorliegende Dissertation umfasst drei Teile, welche in diesem Manuskript in einer hierarchischen Abfolge präsentiert werden, um die zunehmende Organisationsstufe im untersuchten System wiederzuspiegeln. Die Genregulation wurde zuerst auf der Stufe des einzelnen Gens, dann auf der Stufe wechselwirkender Transkriptionseinheiten, und schliesslich innerhalb kleiner regulatorischer Netzwerke untersucht.

Der erste Teil dieser Arbeit konzentriert sich auf einzelne Transkriptionseinheiten regulatorischer Netzwerker und die Mechanismen, welche ihrer Kontrolle zu Grunde liegen. Promotorsequenzen können als Kontrollbausteine von Genen betrachtet werden, welche die Transkription modulieren, indem sie unterschiedliche zelluläre Signale integrieren. Ihre Aktivität kann mit Hilfe mathematischer Werkzeuge beschrieben werden, welche durch generelle Konzepte aus den unterschiedlichen Bereichen von Wissenschaft und Ingenieurwesen inspiriert sind. Unser Ansatz ist eine iterative, auf experimentellen Daten basierte, mathematische Modellierung mit Promotor Bibliotheken synthetischer Bindungsstellen für ausgewählte transkriptionelle Aktivatoren unterschiedlicher Affinität. Diese Bibliotheken wurden für *in vivo* Experimente mit der Hefe *Saccharomyces cerevisiae* verwendet, um unsere theoretischen Modelle zu verbessern. Das Hauptziel dieses Teils der Arbeit war die Bestimmung der intrinsischen Eigenschaften eines Gensystems, wie Kooperativität und Bindingsaffinität von Transkriptionsfaktoren an seine regulatorischen Sequenzen. Die entwickelte Methode bietet eine neuartige Möglichkeit um Genregulation über ein breites Aktivitätsspektrum *in vivo* zu studieren ohne den Zellstoffwechsel zu stören. Wir konnten das Aktivierungsprofil des *GAL1* Promotors, welcher ein Modellsystem für Genregulation ist, auflösen. Der *GAL1* Promotor zeigte hohe Reaktivität bei einem niedrigen Induktionsniveau aber eine graduale Antwort bei höherer Aktivität. Mehrere Aktivatorbindungsstellen führen zu einer mehr-als-additiven Erhöhung der Genexpression im Vergleich zu Promotoren mit einer einzigen Bindungsstelle. Diese Resultate zeigen, dass Transkriptionsaktivatoren nicht direkt, sondern über Komponenten der Transkriptionsmaschinerie kooperieren können.

Im zweiten Teil dieser Arbeit wurde unser Ansatz auf wechselwirkende transkriptionelle Systeme erweitert. Transkriptionelle Interferenz wird definiert als der inhibierende Einfluss eines transkriptionellen Prozesses, direkt und *in cis*, auf einen anderen transkriptionellen Prozess. Dieser Mechanismus der Regulierung bedingt eine antagonistische Beziehung zwischen Transkriptionseinheiten und findet sowohl in der Hefe, als auch in höheren Eukaroyoten statt. In der transkriptionellen Interferenz werden Signale der beiden antagonistischen Aktivatoren durch das Tracking der Polymerase entlang der DNA kombiniert. Die Grundlagen dieser Signalübertragung und wie diese Signale zur Kontrolle der Genexpression verwendet werden, wurden untersucht, indem ein doppeltes Kontrollsystem in der Hefe entwickelt wurde, das mittels equilibrium und non-equilibrium Modellen der Transkription interpretiert wurde. Antagonismus, welcher durch die Bindung eines Aktivators vor der TATA-box erzeugt wird, zeigt das charakteristische kompetitive Inhibitionsmuster, während eine Bindung nach der TATA-box eine nicht-kompetitive Hemmung hervorruft. Unter niedrigen Induktionsbedingungen wurden positive Effekte beobachtet, welche durch Kooperativität zwischen Transkriptionsfaktoren erklärt werden können.

Im dritten Teil wurde der Prozess der Genregulation innerhalb kleiner regulatorischer Netzwerke untersucht, wobei besondere Aufmerksamkeit auf Rückkopplungsschleifen im regulatorischen Netzwerk des Galaktose-Metabolismus gelegt wurde. Basierend auf dem erlangten Wissen aus den vorangehenden Teilen dieser Arbeit wurden die Eigenschaften von Promotoren weiter untersucht, um den Einfluss intrazellulärer Parameter auf quervernetzte genetische Schaltkreise aufzuklären. Die Verwendung von synthetischen, in Hefezellen eingeführte genetischen Schaltkreisen, zusammen mit stochastischer Modellierung, erlaubte uns den Einfluss der Parameter, welche die Dynamik der Genexpression beeinflussen, effizienter zu interpretieren. Ferner konnten wir mit diesem Ansatz Aspekte der physiologischen Anpassung der Hefe in einem molekularevolutionsbiologischen Rahmen besser verstehen. Promotoren, welche durch divergente Evolution entstanden sind, wie der *GAL1*- and *GAL3*-Promotor, haben ihre Funktion unterschiedlich optimiert. Das Gen, welches bei Nährstoffmessung beteiligt ist, zeigte eine schnelle Antwort auf neue Umgebungsbedingungen. Im Gegensatz dazu speichert das Gen, welches im metabolischen Teil des GAL-Netzwerkes involviert ist, vorherige Bedingungen (effizienter), in dem es Kooperativität der Aktivatorenbindung und eine reduzierte basale Expression aufweist.

Contents

I Systematic Study on Promoter Activation   **1**

1 Synopsis   **2**
1.1 Introduction 3
1.2 Method 5
1.3 Results 7
    1.3.1 Proof of Concept: rtTA–*tet*-operators Interaction 7
        1.3.1.1 Calibration: Single *tet*-operator Library 7
        1.3.1.2 Measurement: Multiple *tet*-operator Library 7
        1.3.1.3 Equivalence Between Variables *in vivo* 8
    1.3.2 Application of the Promoter Library Approach to the *GAL1* Promoter 11
        1.3.2.1 Calibration: Single *Gal4pBS* Library 11
        1.3.2.2 Measurement: Multiple *Gal4pBS* Library 11
        1.3.2.3 Dynamical Range of Activation and Sensitivity 12
1.4 Discussion 15
1.5 Conclusions 15

2 Introduction   **17**
2.1 Background 17
2.2 Graded *versus* Switch-like Regulation in Genetic Systems 17
2.3 *Saccharomyces cerevisiae* as a Model Organism to Study Transcription 20
2.4 The Transcription Machinery of *Saccharomyces cerevisiae* 20
2.5 Ligand Binding & Enzyme Kinetics 22
2.6 Mathematical Description of Gene Regulation 23
2.7 Promoter Engineering to Study Gene Regulation 23
2.8 Data Analysis & Mathematical Tools 24
    2.8.1 Saturation Functions for Single Activator Binding 24
        2.8.1.1 Rapid-Equilibrium Assumption to Derive Saturation Functions 24
        2.8.1.2 Steady-State Assumption to Derive Saturation Functions 27
    2.8.2 Adair Equation for Multiple Activators Binding 30
        2.8.2.1 General Adair Equation for Enzyme Kinetics 31
            2.8.2.1.1 Example: Hemoglobin Modeling with Adair Equation 34
        2.8.2.2 Adaptation of the Adair Equation to Gene Regulation . . 36
    2.8.3 The Hill Equation 38
    2.8.4 Logarithmic Sensitivity 40
    2.8.5 Parametric Representation of Gene Expression Data 42
    2.8.6 Logistic Function 44

3 Material and Methods — 48

3.1 Methods — 48
3.1.1 Data Analysis — 48
- 3.1.1.1 Flow Cytometry — 48
- 3.1.1.2 β-Galactosidase CPRG assay — 48
- 3.1.1.3 Real Time PCR — 49

3.1.2 Molecular Biology Methods — 49
- 3.1.2.1 Yeast Transformation (Lithium Acetate Method) — 49
- 3.1.2.2 Bacteria Growth — 49
- 3.1.2.3 Yeast Growth — 49
- 3.1.2.4 Molecular Cloning — 49
- 3.1.2.5 Promoter Libraries — 50
  - 3.1.2.5.1 *tet*-Operator Library in *EGT2* Core Promoter — 50
  - 3.1.2.5.2 Gal4pBS Library in *GAL1* Core Promoter — 50
- 3.1.2.6 Activators — 50
  - 3.1.2.6.1 rtTA — 50
  - 3.1.2.6.2 GEV — 51
  - 3.1.2.6.3 tetR-Ssn6 — 51
  - 3.1.2.6.4 rtetR-Ace2p — 51
  - 3.1.2.6.5 rtetR-Swi5p — 51
  - 3.1.2.6.6 rtetR-NLS-Swi5pABCD — 51

3.2 Materials — 52
3.2.1 Data Analysis — 52
- 3.2.1.1 Flow Cytometry — 52
- 3.2.1.2 β-Galactosidase CPRG assay — 52
- 3.2.1.3 Real Time PCR — 52

3.2.2 Molecular Biology Methods — 52
- 3.2.2.1 Yeast Transformation — 52
- 3.2.2.2 Plasmid Isolation from *E. coli* cultures — 53
- 3.2.2.3 Inducer Stock Solutions — 53
- 3.2.2.4 Bacteria Strains — 53
- 3.2.2.5 Yeast Strains — 53

4 Results — 55

4.1 Context-Dependent Effects on Gene Regulation — 55
4.1.1 Introduction — 55
4.1.2 System Description — 56
- 4.1.2.1 Transcriptional Activators — 56
- 4.1.2.2 Promoters — 57

4.1.3 Results and Discussion — 58
4.1.4 Conclusions — 60

4.2 Systematic Study on the Effect of Multiple Activator Binding Sites in Promoters — 65
4.2.1 Introduction — 65
4.2.2 Full-Synthetic Systems — 66
- 4.2.2.1 System Description — 67

		4.2.2.2	Data Interpretation Using the Hill Function	68
		4.2.2.3	Data Interpretation Using the Adair Equation	75
			4.2.2.3.1 Sequential Non-linear Curve Fitting	75
			4.2.2.3.2 Global Non-linear Curve Fitting of Single Datasets	83
		4.2.2.4	Discussion	92
	4.2.3	Semi-Synthetic Systems		93
		4.2.3.1	System Description	94
		4.2.3.2	Results	95
		4.2.3.3	Discussion	96
	4.2.4	Natural Systems		99
		4.2.4.1	System Description	100
		4.2.4.2	Testing the Natural Systems Approach: rtTA–tet-operators Interaction	102
			4.2.4.2.1 Calibration: Single tet-operator Library	102
			4.2.4.2.2 Measurement: Multiple tet-operator Library . .	104
			4.2.4.2.3 Proof of Concept	106
		4.2.4.3	Application of the Natural System Approach to $GAL1$ Promoter Activation	108
			4.2.4.3.1 Calibration: Single $Gal4pBS$ Library	108
			4.2.4.3.2 Measurement: Multiple $Gal4pBS$ Library	110
		4.2.4.4	Discussion	112

5 Conclusions and Future Work 117
5.1 Conclusions 117
5.2 Future Work 120

6 Appendix I 122
6.1 Promoter Sequences 122
    6.1.1 $P_{tetO2inSIC1}$ 122
    6.1.2 $P_{tetO5inEGT2}$ 122
    6.1.3 $P_{tetO6inEGT2}$ 122
    6.1.4 $P_{tetO1inCYC1}$ 123
    6.1.5 $P_{tetO2inCYC1}$ 123
    6.1.6 $P_{tetO7inCYC1}$ 124
    6.1.7 $P_{tetOinEGT2}$ 124
    6.1.8 $GAL1$ Promoter 125
6.2 tet-Operators 126
6.3 Gal4p Operators 128
6.4 Comparison pAnt117 (pAnt27-derived (no $FIG1$-integration-locus)) and pAnt156 (pPR173-derived (with $FIG1$-integration-locus)) 131
6.5 Positional Effect of $Gal4pBS$ in $GAL1$ Promoter 132
6.6 Feedback of $GAL1$ Gene 133

7 Appendix II 135

II Transcriptional Interference: Interaction between Transcriptional Processes 139

CONTENTS

1 Introduction to Transcriptional Interference   **140**

III   The Role of Promoters within Small Regulatory Networks   184

1 Introduction   **185**

2 Method   **186**
  2.1 Experimental Methods 186
     2.1.1 Promoters 186
     2.1.2 Basal Expression Measurement 187
     2.1.3 Adaptation 187
  2.2 Stochastic Simulations 187

3 Results and Discussion   **189**

4 Conclusions   **191**

Overall Conclusions   **193**

Acknowledgements   **195**
  General Acknowledgements 195
  Part I 195
  Part II 195
  Part III 195

Glossary   **197**

Bibliography   **201**

Part I
Systematic Study on Promoter Activation

1 Synopsis

The first part of this manuscript represents ongoing work currently involving many persons and potentially subject to further developments. Hence, this first brief chapter called 'Synopsis' summarizes most important aims and achievements related to the tools employed to describe gene regulation with help of synthetic promoter libraries of variable affinity for transcriptional regulators. The document contains relevant figures and short description of the applied methodology. However, people with interest in the application or further developments of the work currently reported in the following document can find more detailed and contextualized information in the main text of the manuscript.

Abstract

Understanding and predicting how promoter sequences affect transcriptional activity is crucial to understand the complexity of gene regulation. The reaction of genes to changing conditions occurs through transcriptional regulators that mediate signals to promoters upon binding to specific DNA sequences (regulator binding sites). Graded and switch-like responses are fundamental aspects of signal transduction representing connections between genetic units. The role of multiple regulator binding sites in promoter sequences is still unknown. Multiple binding sites and cooperative effects between regulators potentially play a role in promoter mediated signal transduction. This would provide an efficient signal conversion mechanism embedded in the promoter sequence to achieve graded and switch-like response patterns. To address this question we have used a modern genetic approach relying on promoter engineering. Synthetic promoter libraries of variable affinity for selected transcriptional regulators have been used to perform *in vivo* experiments and results were analyzed with mathematical tools widely applied in science and engineering. The *GAL1* promoter activation could be accessed under minimally invasive experimental conditions over a broad range of activity. *GAL1* promoter displayed high reactivity at low induction but graded response when exposed to higher induction levels. Multiple binding sites contributed a more-than-additive increase of expression compared to the promoter activated by a single binding site revealing that activators can cooperate not by direct interaction but through components of the transcriptional machinery. The methodology proposed in this work provides a new way to study the regulation of genes *in vivo* over a broad range of activities with minimally invasive impact on cellular metabolism. The advantages of a genetic approach accounting for *in vivo* context specificity are combined here with the broad, continuous range of accessibility characteristic of comparable *in vitro* studies.

1.1 Introduction

The complexity of living organisms implies highly sophisticated mechanisms of gene regulation in order to achieve the correct functioning of cellular processes. Transcription is the primary regulatory process used by cells to control the performance of genetic circuits and it is mainly determined by the interplay between molecular components such as transcription factors and DNA regulatory sequences. Simple pathway structures can generate a variety of responses not only depending on the strength of the stimulus, but also on the processing unit responsible for the input-to-output signal conversion. A crucial point in studying transcriptional regulation is thereby to understand and predict how the architecture of promoters affects transcriptional activity [1]. Consequently, the measurement of intracellular parameters describing the process at the level of regulator binding to gene regulatory sequences, like cooperativity coefficients and interaction strength, is required for the analysis of the behavior of molecular systems in living cells.

Graded and switch-like responses are fundamental aspects of signal transduction since they represent major classes of connections between genetic units [2, 3, 4, 5, 6, 7, 8, 9]. Both ways of integrating signals are required in different conditions: Switch-like response may be relevant for ultrasensitive pathways while graded response is necessary to provide a quantitatively adequate reaction to a certain stimulus. Graded responses are for example observed in inflammatory reaction induced by microbes, in which response is proportional to the microbial load as well as in the graded expression of p53 as a consequence of different intensities of ionizing radiation [10, 11]. Examples of switch-like responses are found in the process of *Xenopus* oocytes maturation [12] or in the

1.1 Introduction

mechanism controlling the genetic switch that determines lytic or lysogenic life cycle of the bacteriophage λ [13, 14, 15, 16].
It is commonly assumed that both in prokaryotes and eukaryotes, single gene elements are intrinsically dominated by graded response type of regulation and that modes of connectivity in their network context (e.g. positive feedback) can generate all-or-none gene expression patterns [5, 17, 12, 18]. It is also known that certain transcriptional regulators can bind different regulatory sequences in the genome in order to coordinate activity of several genes. Interestingly, the promoter sequence of these genes regulated in concert often have variable number of binding sites for their common regulator. Pleiotropy is a widespread regulatory feature among organisms and combined with the presence of promoter sequences regulated by multiple binding sites hints for a potentially relevant role played by synergic effects between regulators. In enzyme kinetics and ligand binding, multiple ligand binding to receptors is commonly associated with binding cooperativity and thus with switch-like response patterns while single ligand binding with graded responses. Hence, regulator binding cooperativity could provide additional means for modulating gene activation without requiring mechanisms of positive retroaction. Accordingly, same signal could be decoded differently by different genes by simply embedding in their promoter architecture a signal conversion mechanism supporting from single to multiple regulator binding and thereby modulating their activation profile with different degree of sensitivity from graded, dose-dependent induction to more switch-like to binary response.

In order to understand function and properties of genes regulated by promoters containing single *versus* multiple activator binding sites and to shed light on their corresponding activation profile, we developed a genetic system using modern techniques based on promoter libraries containing synthetic binding site sequences of variable affinities for selected transcriptional activators. The experimental findings have subsequently been interpreted using mathematical modeling inspired by general concepts widely applied in different fields of science and engineering.

Gene activity can be efficiently described using *saturation functions* in order to map degree of induction (input) to gene expression level (output). Saturation functions are transfer functions that relate input and output in biochemical systems providing a measure for the relationship between substrate concentration and rate in enzyme kinetics or fraction of complex formation in molecular association studies as a function of the concentration of one system's component.

Experimentally, protein-DNA binding is classically studied with *in vitro* techniques consisting in applying gradually increasing regulator concentration to purified DNA fraction and monitoring protein-DNA complex formation. However, even if saturation functions retrieved in this way cover broad activation ranges, they often fail to represent the actual process occurring *in vivo* due to the absence of cellular components present under natural conditions. Saturation functions are alternatively evaluated *in vivo* by sampling a continuum of gene expression through few discrete datapoints anyway providing an incomplete picture of the gene activation process [19].

Gradual variations of transcription factor concentration cannot be achieved *in vivo* in most of the cases and whenever possible it often implies important consequences for cellular physiology. The proposed technique consists instead in varying binding site affinity by constant regulator concentration and provides continuous access of saturation functions allowing to take into consideration genetic and cellular aspects relevant for the description of the studied system. This method combines the advantages of classical genetic approaches commonly used in molecular and cellular biology with the broad activity range usually accessed in *in vitro* experiments.

1.2 Method

The method presented in this work allows to get insights in the natural process of gene regulation with minimal perturbations affecting cellular physiology. While the endogenous copy of the gene under study executes its function unperturbed, an additional copy of its promoter reports in parallel its activation state through the expression of a reporter gene. Further, in order to obtain saturation functions characterizing the gene activation profile (instead of single data points), a series of identical promoters differing *only* in the affinity of the regulator binding site sequence is introduced in different cells of the same strain. This promoter library of variable affinity is obtained by introducing gradual mutations in the consensus binding site sequence for a given transcription factor and can be obtained by modern techniques of *in vitro* and *in vivo* directed evolution [20, 19, 21, 22].

By exposing cells containing such promoter library of variable affinity to the same conditions (i.e., setting a constant regulator concentration), the same information is retrieved as by varying the concentration of a regulator binding a promoter of constant affinity. This statement is supported by the mathematical formalism developed for the description of chemical binding processes where ligand's affinity and concentration are treated as equivalent variables. Reaction kinetics theory assumes in fact that both regulator concentration and affinity compensate each other.

The methodology presented here is decomposed in two parts, called *calibration* and *measurement*. The goal of the calibration phase is to determine individual contributions that single binding site sequences bring to system's activation. Applying different mutations to consensus sequences, the binding affinity for a particular regulator can be modulated yet conserving specificity. Once the system calibrated, thus knowing the extent of system activation that every individual mutant binding site in the library contributes to, the measurement phase consists of introducing for each mutant sequence used in the single operator library, its corresponding construct composed of the same affinity mutant present in multiple copies (constituting the multiple operator library).

Correlating parametrically single with multiple operator libraries allows to access the saturation function that describes the gain in promoter activation realized upon the presence of multiple binding sites in the regulatory sequence under study (see Fig.A1).

1.2 Method

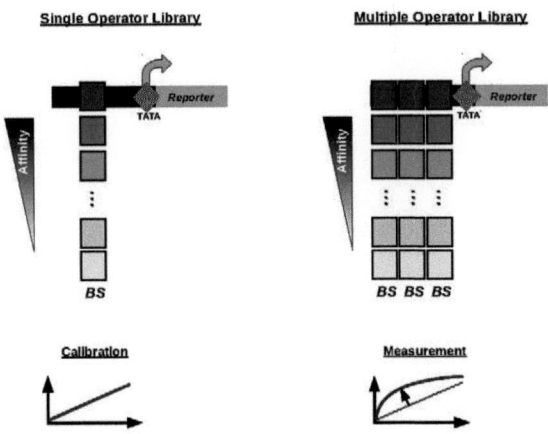

Figure A1: Promoter libraries of variable affinity. The method consists in creating two libraries containing mutant sequences of variable affinity for a given regulator. The single operator library is realized by applying mutations in the activator consensus sequence that affect promoter activation in a graded way. The multiple operator library uses the same mutants as in the single operator library, with the difference of being present in blocks of repetitive units within the studied promoter. Calibration of the system is achieved by using the single operator library while the measurement phase is carried out using the multiple operator library: Under same external conditions, the deviation in promoter activation between single and multiple operator libraries reflects the gain represented by the presence of multiple activator binding.

1.3 Results

1.3.1 Proof of Concept: rtTA–*tet*-operators Interaction

The approach was initially tested to study interaction between rtTA and *tet*-operators in the core promoter of the *EGT2* gene. Representing an exception among transcriptional factors, doxycycline tunable rtTA–*tet*-operators interaction offers a unique opportunity to externally fix a certain amount of active regulator in living cells and consequently allows to study the effect of binding site mutations under different activator concentrations. This proof of concept allowed us to ascertain the validity of the technique on an easily controllable system well characterized by mutational studies found in the literature. However, the major advantage of this system consists of the possibility to access saturation curves retrieved by gradually varying operator affinity, and to compare the outcome with the corresponding activation profile obtained by graded doxycycline induction.

1.3.1.1 Calibration: Single *tet*-operator Library

The *tet*-operator (*tetO1* type) is a 19 bp palindromic sequence (see Fig.A2 (Top center)) to which the rtTA homodimer binds. The advantage for our study relies on the symmetry of these binding partners at the molecular level. The binding site sequence can be altered by single nucleotide mutations in one half of the operator only, resulting in conserved specificity on one side and altering affinity for rtTA on the other side.

Mutagenesis studies have been performed in *Escherichia coli* and the effect of every nucleotide mutation on TetR binding has been reported in [23]. Based on this information, we synthetically obtained different single mutant operator sequences that were inserted upstream of the core promoter of the *EGT2* gene (-193 from start site), transformed into wild type *Saccharomyces cerevisiae* (W303A), mated with an analogous wild type strain (W303α) containing P_{CLN3} (cycline 3 promoter) driven rtTA, and expression of diploids was measured by flow cytometry under high inducing conditions (20 µM doxycycline). The basal activity due to the *EGT2* core promoter was roughly 2-fold higher in yeast than in the bacterial system (8% in *Escherichia coli* ([23]) and 17% in yeast with respect to the wild type). However the outcome showed that both organisms expressed highly correlated activities (correlation coefficient > 0.99) in relation to the same mutations (see Fig.A2 (Upper left panel)).

1.3.1.2 Measurement: Multiple *tet*-operator Library

After assessment of how single mutations in the *tet*-operator influence gene activation individually, the next step consisted of testing the same mutant sequences in corresponding promoter systems containing multiples of their copies. Hence, we synthetically introduced multiple operator repeats consisting of four identical *tet*-operator sequences spaced by 6 or 8 bp in the core promoter of the *EGT2* gene (-193 from start site). The resulting vectors were subsequently transformed in wild type strains of *Saccharomyces cerevisiae* (W303α) containing P_{CLN3} driven rtTA. Similarly, the corresponding single *tet*-operator library has also been transformed in the wild type W303α strains containing P_{CLN3} driven rtTA and both resulting series of haploid strains have been measured at the same time by flow cytometry at constant doxycycline concentrations (20 µM or 6 µM). When expressing multiple and single operator systems as a function of the same induction it is convenient to combine the two corresponding responses and represent the deviations between each other parametrically. Since the two promoter libraries contain different numbers of the same binding site sequence, their activity can be combined as a function of variable affinity. Correlating the multiple operator library (on the y-axis) with the single operator library (on the x-axis), allows characterization of the enrichment in gene activity due to multiple binding sites as a function of the single binding site driven promoter. A reference can be constructed by correlating single *versus* single operator library resulting in a diagonal identity function accounting for zero enrichment. Therefore, as predicted theoretically, exposing cells containing libraries of mutant *tet*-operators in different number resulted in the expected distortion of the calibration curve (Upper right panel *versus* Lower right panel of Fig.A2). Also under different induction conditions (i.e., setting different activator concentrations through doxycycline), the system responded according to expectations by displacing the experimentally accessible region of the same curve to different expression ranges (compare red and black curves of Upper right panel and Lower right panel of Fig.A2).

1.3.1.3 Equivalence Between Variables *in vivo*

The mutant *tet*-operator library could be used further to check the same principle described above in an alternative way. Since activator's affinity and concentration are considered dependent variables, the validity of the methodology presented above where activator's binding site affinity has been varied at constant concentrations, can also be proved equivalently by varying activator concentration at fixed affinities. So, two pairs of mutants (from single and multiple operator libraries), WT and $5T$, were gradually induced by doxycycline and correlated parametrically resulting in the curves shown in Fig.A2 (Lower left panel) together with the same dataset shown in Fig.A2 (Upper right panel). The expression of wild type single operator at 20 µM doxycycline was used as reference to combine the two different sets.

According to theoretical predictions stated by reaction kinetics theory, the equivalence of variables could be verified experimentally supporting the methodology of promoter libraries as an efficient tool to study multiple binding site promoters.

1.3 Results

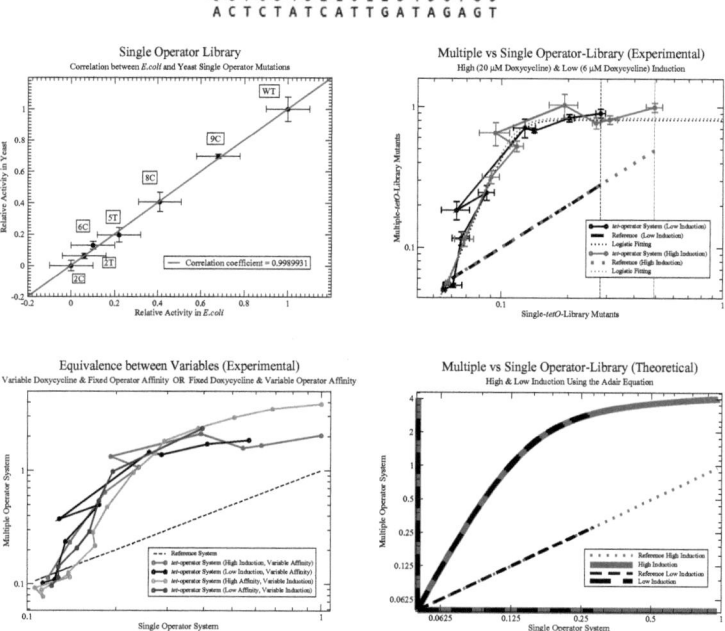

Figure A2: <u>Top center</u>: The palindromic *tet*-operator sequence (*tetO1* type) with positions indicated with positive numbers referring to one half and negative numbers referring to the other half of the sequence. <u>Upper left panel</u>: Calibration phase of single *tet*-operator library. Correlation of single operator library measured in *Escherichia coli* (according to expression values in [23]) and in *Saccharomyces cerevisiae* at high inducing conditions (20 µM doxycycline). Single nucleotide mutations are indicated by boxes, where numbers designate the position according to nucleotide coordinates described in the upper sequence, and letters correspond to substitutive nucleotides at that position. The basal expression value is 8% and 17% of the wild type sequence expression in *Escherichia coli* and *Saccharomyces cerevisiae*, respectively. The correlation coefficient is > 0.99, reproducibility of LacZ expression measurements in bacteria lie within ±10% ([23]) and error bars of yeast's expression measurements show standard deviations from at least 3 replicates. <u>Upper right panel</u>: Measurement phase of multiple *tet*-operator library. The x-axis represents the range of expression covered by the single operator library while on the y-axis by the multiple operator library (consisting of a block of four *tet*-operators). Measurement curves of the multiple operator library corresponding to high (red lines, 20 µM doxycycline) and low (black lines, 6 µM doxycycline) induction are obtained by combining expression values of x and y-coordinates while reference systems are constructed by the same procedure but comparing single *versus* single operator library, thus giving a diagonal line (thick, dashed and dotted lines). Error bars are calculated with standard error of the mean from three experiments and expression values are normalized to their maximal value. Thin dashed

1.3 Results

lines indicate the range of activity of single operator library at high (red) and low (black) induction. Thin dotted lines are fitted functions to the datasets (standard logistic function ($y = \frac{A0}{A1+e^{-A2 \cdot x}} + A3$) where the resulting parameters are: $A0 = 1.59 \cdot 10^{-3}$; $A1 = 1.96 \cdot 10^{-3}$; $A2 = 63.28$; $A3 = 2.60 \cdot 10^{-9}$ for the low induction and $A0 = 1.38 \cdot 10^{-3}$; $A1 = 1.65 \cdot 10^{-3}$; $A2 = 66.31$; $A3 = 5.40 \cdot 10^{-11}$ for the high induction curve. Lower right panel: Theoretical saturation curves in parametric representation. Multiple operator curves are generated using the classical Adair equation ([24]) with four activator binding sites ($n = 4$), the association equilibrium constant $K = 1$, and cooperative enhancement factors $a = 10$, $b = c = 1$. Similarly, single operator system has been simulated with $n = 1$, $K = 1$. In addition, basal expression was set at 0.05 units and added to the respective equations. Induction was arbitrarily set at 10 and 0.3 corresponding to the denomination high and low induction, respectively. On the x-axis it is represented the range of expression covered by the single operator system while on y-axis by the multiple operator system. Curves corresponding to high (red lines) and low (black lines) induction are obtained by combining expression values of x and y-coordinates while reference systems are constructed by the same procedure but comparing single *versus* single operator system, thus giving a diagonal line (thin dotted and dashed lines). Lower left panel: Equivalent variable compensation in rtTA–*tet*-operator system. Two sets of curves generated by the same system are compared. In the first set (same set presented in the Upper right panel), operator affinity was varied gradually at two different activator concentration levels: Single and multiple operator libraries were exposed to high (20 µM doxycycline) and low (6 µM doxycycline) induction (red and black curves, respectively) and their respective expression correlated parametrically using binding site affinity as common parameter. In the second set, operator affinity was kept constant at two different levels and the measurement carried out under graded doxycycline induction: Single and multiple operator constructs containing WT (high affinity, wild type (100% activity)) and $5T$ (low affinity, 22% of wild type activity) operators (green and blue curves, respectively) were exposed to the same doxycycline gradient and correlated parametrically using doxycycline concentration as common parameter. To combine the two different sets the expression of wild type single operator at 20 µM doxycycline was used as reference. The experiments were obtained with strains and plasmids detailed in Tables 1, 2, 3.

1.3.2 Application of the Promoter Library Approach to the *GAL1* Promoter

After having assessed validity of the experimental design on the rtTA–*tet*-operator test system, we further extended the study by applying the approach to the *GAL1* promoter. The *GAL1* promoter (P_{GAL1}) controls the expression of the galactokinase involved in the metabolism of galactose in yeast and contains four binding sites (*Gal4pBS*) at close distance for the transcriptional activator Gal4p.

1.3.2.1 Calibration: Single *Gal4pBS* Library

Gal4p activator binds several target genes in the yeast genome, each differing in number and sequence of binding sites. The target binding sites are characterized by the consensus sequence $CGG[N]_{11}CCG$ where N represents unspecified nucleotides. It allows to maintain binding specificity and at the same time can display variable range of affinities for Gal4p conferred by the composition of the unspecified nucleotides. Moreover, purine-pyrimidine conversion ($CGG[N]_{11}CCG \rightarrow GCC[N]_{11}GGC$) abolishes Gal4p binding and can be used to create the single operator library by mutating three out of the four *Gal4pBS* of P_{GAL1}.

All operator variants known to specifically bind Gal4p were individually analyzed under high inducing conditions (0.5% galactose and 2% raffinose) in order to identify candidates best suited for being employed in the measurement phase of the study (see Fig.A3 (Upper left panel)). The screening outcome revealed that the vast majority of native binding sites display relatively strong and narrow range of affinities and that an important decrease in affinity could only be attained by mutating the first nucleotide in the consensus sequence. The screening was thereby extended to more candidates and a single operator library of logarithmically graded expression constructs could be constituted and employed for the measurement phase.

1.3.2.2 Measurement: Multiple *Gal4pBS* Library

The *Gal4pBS* selected during the calibration phase were used to replace the four binding sites of P_{GAL1}. The expression range of all the components constituting single and multiple operator libraries was expressed as a function of the wild type P_{GAL1} expression level at high inducing conditions (0.5% galactose and 2% raffinose). Afterward, multiple *versus* single operators have been correlated through parametric representation together with the corresponding reference system and the wild type P_{GAL1} exposed to identical conditions (see Fig.A3 (Upper right panel)).

The contribution of multiple activator binding sites in P_{GAL1} activation was predominantly consisting of a more-than-additive increase in absolute expression level with respect to single binding sites. Consistent with the obtained saturation curve, the high induction conditions applied to the *GAL1* promoter library were reflected by wild type P_{GAL1} activation exposed to the same conditions (blue star in Fig.A3).

1.3.2.3 Dynamical Range of Activation and Sensitivity

In order to compare the two analyzed systems, their corresponding dynamical ranges of activation with respect to single operator systems have been represented in the same figure (see Fig.A3 (Lower left panel)). In addition, their datasets expressed parametrically were fitted to a standard four-parameter logistic equation (see black and red dotted lines in Fig.A2 (Upper right panel) and green dotted line in Fig.A3 (Upper right panel)) and the resulting optimally fitting functions used to compute logarithmic sensitivities represented graphically as a function of single operator activation (see Fig.A3 (Lower right panel)).

The results showed that the *GAL1* promoter is very reactive to low induction levels, but monotonously increasing its response in a proportionate manner at higher activity. The system covers a broad range of induction exhibiting graded response around its highest dynamical range of activation situated at roughly 10-fold higher level with respect to the single operator system. The effect of multiple operators in *GAL1* promoter resulted in rising system's response more homogeneously compared to the *tet*-operator test system. The *tet*-operator system displayed indeed a much sharper activation profile and was confined to a limited excursion range with narrower sensitivity peak.

1.3 Results

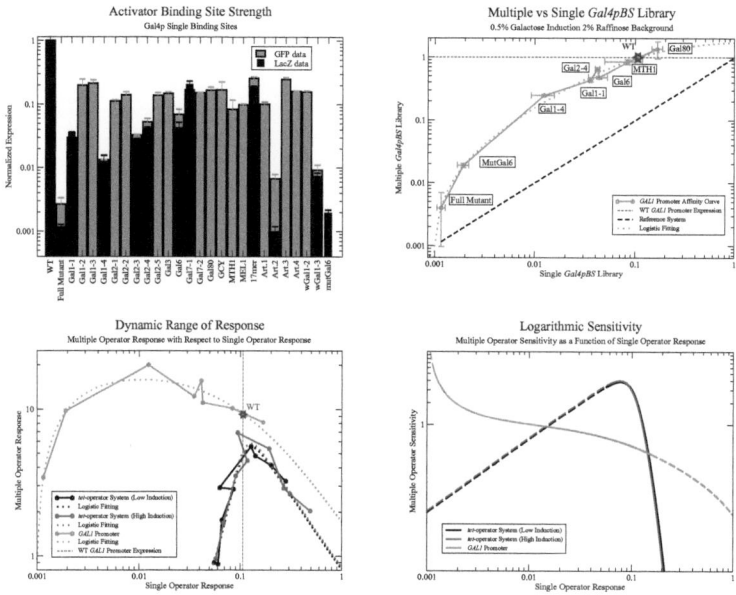

Figure A3: Upper left panel: Gal4p binding sites affinity. Different variants of the Gal4p consensus sequence ($CGG[N]_{11}CCG$) were cloned in the fourth, proximal position of P_{GAL1} and the corresponding activities have been measured under high inducing conditions (0.5% galactose and 2% raffinose) by flow cytometry (green bars) or β-Galactosidase CPRG assay (black bars) depending on the expression range. Expression was normalized with respect to wild type P_{GAL1} expression (WT), full mutant indicates a promoter variant in which all four binding sites were mutated according to purine-pyrimidine conversion ($CGG[N]_{11}CCG \rightarrow GCC[N]_{11}GGC$) and error bars show standard deviations from two independent experiments comprising each at least 3 replicates. Native binding site sequences are denoted by the name of the gene containing them in its promoter sequence and in case a promoter contains more Gal4p binding sites then they are numbered. 17mer corresponds to a high affinity artificial version of the consensus sequence (according to [25]) and binding sites situated to its right in the bar plot are artificial operators found in the literature ([26, 27]). The last sequence on the right side of the bar plot corresponds to the native binding site of the $GAL6$ gene with a single nucleotide mutation in the first position ($C \rightarrow A$). Upper right panel: Measurement phase of multiple operator library. The x-axis represents the range of expression covered by the single operator library while on y-axis by the multiple operator library. Measurement curves of the multiple operator library are obtained by combining expression values of x and y-coordinates under GAL genes inducing conditions while reference systems are constructed by the same procedure but comparing single *versus* single operator library (black dashed line). Blue dashed line indicates the activity level of the wild type promoter under same induction conditions as the promoter libraries. Green dotted line represents a fitted function

1.3 Results

to the dataset (standard logistic function ($y = \frac{A0}{A1+e^{-A2 \cdot x}} + A3$) where the resulting parameters are: $A0 = -0.37$; $A1 = -0.79$; $A2 = -2.27$; $A3 = 1.76$). Lower left panel: Activation profile of *GAL1* promoter together with *tet*-operator system. The effect of multiple operators (four in both cases) is represented with respect to corresponding single operator response. The dynamic range of activation with respect to single operator systems has been computed by dividing datapoint values and fitted functions (from Fig.A2 (Upper right panel) and Fig.A3 (Upper right panel)) by the reference system. The blue star indicates the level of expression corresponding to the wild type *GAL1* promoter exposed to same inducing conditions as the *GAL1* promoter library. Lower right panel: The curves fitted to parametric datasets have been used to compute logarithmic sensitivities of the two systems as a function of single operator response. Continuous lines indicate experimentally accessible range, dotted lines are extrapolated based on the fitted function. The experiments were obtained with strains and plasmids detailed in Tables 4, 5, 6.

1.4 Discussion

The approach presented in this work has been initially validated using a synthetic test system consisting of rtTA–*tet*-operators in order to assess the reliability of the technique under experimental conditions allowing external control. The technique was explored under different aspects and theoretical concepts could be verified experimentally. Subsequently, the technique was applied to the Gal4p-*Gal4pBS* interaction in the *GAL1* promoter. This system represents a model for gene regulation in which external inducers cannot be efficiently used to study promoter activation. Here, promoter activation could be accessed under minimally invasive experimental conditions in living cells over a broad range of activity.

GAL1 promoter was highly reactive at low induction and displayed a graded response at higher activity. Multiple activator binding sites contributed in a more-than-additive increase of expression over a broad range of induction compared to the promoter activated by a single activator binding site. *GAL1* promoter has switch-like characteristics that allow sensitive reaction to changing conditions at low induction, however conserving the ability to modulate its response proportionally when exposed to higher induction levels.

The role of multiple activator binding sites in the promoter of *GAL1* and possibly in the coregulated genes is mainly responsible for increasing absolute expression. The more-than-additive characteristics of this effect do not rely on binding cooperativity as defined in enzyme kinetics but rather occur at the level of the transcriptional machinery organization. This supports the idea that eukaryotic activators can cooperate not by directly interacting but by simultaneously touching some components of the transcriptional machinery [28, 29].

1.5 Conclusions

The presented methodology combines the advantages of a genetic approach accounting for *in vivo* context specificity and at the same time provides saturation functions over a continuous, broad range of induction. The emerging complexity characterizing gene systems can be efficiently taken into consideration by employing the proposed technique. Activation curves resulting from our study are composite functions combining higher level properties of promoter regulation with respect to simpler *in vitro* studies. The applicability of the technique was confirmed from *E.coli* to yeast and is potentially extendable to higher eukaryotes. Its application could range from the characterization of mechanisms underlying activation of selected genes to a synthetic experimental setup in which precise tuning of gene expression is often required to obtain desired properties in genetic networks. Elements of such promoter libraries could thereby be employed to improve pathway behavior and consequently optimize the design of artificial networks.

2 Introduction

2.1 Background

The human genome consists of 3×10^9 DNA base pairs and is estimated to contain 20,000-25,000 genes. Genes represent only about 2% of the human genome; the remainder consists of non-coding regions, whose functions may include providing chromosomal structural integrity and regulating where, when, and in what quantity proteins are made [30].

Gene regulatory sequences can be seen as control units responsible for the correct functioning of the underlying interactions between genes. Moreover, these sequences have the capacity to integrate a variety of signals induced by cell's state or environment [31, 32, 33, 34, 35]. This usually results in the activation of certain key genes which play a determining role for the cell to react to a given perturbation in an appropriate way. Such cellular responses are more the result of an ensemble of cooperating units which provides a concerted response than due to single isolated genes [31, 36, 35, 37]. The ability of cells to respond to perturbations often results in mechanisms involving physicochemical interactions between molecular species and frequently exhibits complex mechanisms that can display favorable emergent properties in a system [38, 39, 40, 21, 6]. This allows cellular systems to perform complicated tasks such as graded or switch-like response [2, 3, 6, 7, 8, 9], transformation of gradients into discrete changes in gene expression [38], signal propagation modulation [41], sharpening of the spatio-temporal response's delays and oscillations [42, 43, 44, 45], bistability and memory effects [46, 47, 48, 17, 49] which results in an adequate, fine tuned control.

Organismal complexity implies highly sophisticated mechanisms of gene regulation in order to achieve the correct functioning of cellular processes. Transcription is the primary regulatory process used by cells to control the performance of genetic circuits and it is mainly determined by the interplay between molecular components such as transcription factors and DNA regulatory sequences. The measurement of intracellular parameters describing modes of regulation of macromolecular components, like cooperativity coefficients and interaction strength, is required for the analysis of the behavior of molecular systems in living cells. Graded and switch-like responses are a center of interest in our work since they represent major connections between genetic units of an organism and thus are relevant for system description of living cells. The purpose of the present work is to validate new mathematical tools inspired by general concepts widely used in different fields of science and engineering using an iterative experimental-based mathematical modeling approach based on promoter engineering. We therefore construct promoter libraries of variable affinities for transcriptional activators that we subsequently use to perform *in vivo* experiments in order to improve our knowledge about the mode of activation of promoter regulatory regions. The goal of this work is to determine system's intrinsic properties like cooperativity coefficients and binding affinities of transcription factors to DNA regulatory sequences in order to access information about the molecular mechanisms underlying the process of gene regulation in living cells.

2.2 Graded *versus* Switch-like Regulation in Genetic Systems

Simple pathway architectures can generate a variety of responses not only depending on the strength of the stimulus, but also on the processing unit responsible for the input-to-output signal transduction. Two crucial aspects of signal transduction are graded and switch-like response. Namely, both ways of integrating signals are required in different environments: Switch-like response may be relevant for ultrasensitive pathways while graded response is necessary to provide a quantitatively adequate reaction to a certain stimulus. The importance of different signal processing is determining depending on the physiological context the cell is exposed to.

In a graded response, the output of a regulatory pathway is determined proportionally to the applied stimulus. The term 'graded' describes a response that achieves a continuous range of activity from fully on to fully off [5]. Activators can bind to gene regulatory sequences and thereby increase the rate of transcription in each cell in a dose-dependent manner. Graded alterations in gene expression may allow fine-tuned regulation that permit proportionate responses to a stimulus. In the latter case, all cells in a population could respond with similar changes in expression and thereby maintain homogeneity [4]. Graded responses are for example observed in inflammatory response induced by microbes, in which response is proportional to the microbial load as well as in the graded expression of p53 in response to different intensity of ionizing radiation [10, 11]. Graded response can also for example play a role in linearizing transcriptional cascade before saturation. In this case negative feed back increases the linearity (reduces distortion) in gene circuits [50]. This effect results in reducing noise at a wide range of inducer concentrations, similarly to linearization applied to other fields of science like electronics, control theory, and neuroscience [51].

On the other hand, mechanisms for regulatory pathways sensitive to small changes in environmental stimuli are required when fast response to changing conditions is expected. Ultrasensitive response is the underlying signal processing mechanism responsible for the occurrence of all-or-none events. Small variations in the input stimulus could thereby lead to large changes in the response, which is commonly encountered in signal transduction cascades and signal propagation modulation [41, 7, 8, 9]. It is thought that both in prokaryotes and eukaryotes, single gene elements are intrinsically dominated by graded response type of regulation and that modes of connectivity in their network context can generate all-or-none gene expression patterns [5]. Mechanisms such as positive feedback loops are known to promote switch-like responses and their role in biological processes seems to be widely spread [17, 12, 18]. An example of switch-like response based on positive feedback is found in the process of *Xenopus* oocytes maturation [12]. Due to positive feedback embedded in the MAPK signaling cascade, the graded stimulus provided by the hormonal concentration responsible for oocytes maturation is converted into a highly switch-like biological response. The biological effect of this mechanism is advantageous in its natural context: With 50% of the stimulus applied to *Xenopus* oocytes, 50% of them would be able to mature completely. In contrast, without such signal conversion, 50% hormone induction would bring all of the oocytes to incomplete maturation (50% maturation assuming linearly proportional input-to-output relation). So, if in nature transient hormonal stimulus is below the maximal level, such mechanism would still allow a certain amount of oocytes to finalize maturation and continue through further developmental stages.

Another mechanisms capable of generating ultrasensitive response is binding cooperativity. This implies that upon binding of a ligand to its target receptor, the succeeding ligand will bind with a different affinity. Ligands bound to the receptor can thereby enhance or hinder subsequent ligand binding (positive of negative cooperativity, respec-

2.2 Graded versus Switch-like Regulation in Genetic Systems   Introduction

tively). Positively cooperative systems have sigmoidal stimulus/response curves: At very low input signal levels, the response increases almost linearly with the stimulus and only after a certain threshold the input is amplified and output signal undergoes drastic changes. Binding cooperativity is well known in the field of enzyme kinetics and other systems of interacting molecules (e.g. oxygen binding to hemoglobin). However, even if it is expected that some kind of interaction occurs between regulators binding to adjacent DNA sites, either protein-protein based or DNA mediated (e.g. protein induced fit of DNA) [52], the role of such mechanism in the context of gene regulation is still obscure. It is in fact unclear to what extent DNA regulatory sequences alone can convert linear inputs to sigmoidal responses. The potential cooperativity of transcription factor binding to DNA regulatory sequences is especially interesting for the reason that some regulators can bind more target genes in the genome, sometimes differing in the number of binding sites. Binding cooperativity would thus provide means to embed in the gene regulatory sequence itself the capacity to convert the same linear input signal into a graded response through single binding sites or, alternatively, into a switch-like response through multiple binding sites.

An example of binding cooperativity involving genetic circuits is found in the bacteriophage λ in the mechanism of action of the *cro* protein binding to its operators [13, 14]. *Cro*, a repressor protein, works in opposition to the phage's *cI* repressor protein to control the genetic switch that determines whether a lytic or lysogenic cycle will follow infection of host bacteria. This relies on interactions of the *cI* repressor proteins and *cro* repressors which self assemble and bind their operators O_{R1}, O_{R2}, O_{R3}. The *cI* repressor binds with increasing affinity to O_{R1}, O_{R2} and O_{R3}. Adjacently bound *cI* repressors exhibit positive cooperativity [14]. The strength of *cro* dimer binding is strongest to O_{R3}, then O_{R1} and lastly O_{R2}. A competition between the *cro* and repressor proteins ensues, the outcome of which determines whether the phage embarks on a lytic or lysogenic life cycle. Due to positive cooperative DNA binding, *cI* protein binds the operators O_{R1} and O_{R2} thereby constantly driving its own expression and so promoting the lysogenic life cycle and consequently preventing *cro* expression. Upon *cI* degradation (an action initiated by DNA damage and mediated by *recA*), O_{R1} and O_{R2} become vacant derepressing *cro* production. This allows O_{R3} to be bound by *cro* and consequently promoting its own production thus having the effect of switching the life cycle of the bacteriophage λ to the lytic mode [15, 16].

Another example of cooperative binding of regulator to DNA is exemplified by the putative histone acetyltransferase of *Saccharomyces cerevisiae* Spt10p. This transcription factor binds to the upstream activation sequences of histone gene promoters in a cooperative fashion *in vitro* with dissociation constants of $K_D > 1$ µM and $K_D = 45$ nM to single and paired DNA binding sites, respectively. Experimental evidence shows that the DNA binding domain alone is unable to display cooperative binding hinting for protein-protein interaction underlying the observed effect. Moreover, genomic disposition of Spt10p binding sites in the yeast genome and genetic analysis also suggest interaction between Spt10p proteins *in vivo* [53, 54].

These examples illustrate how gene systems have the potential to display sophisticated response patterns based on protein-DNA interactions (transcription factors binding to DNA regulatory sequences) and protein-protein interactions (interactions between transcription factors). However, even if there is evidence proving the presence of cooperativity at the level of regulators binding to promoter sequences, development of new

2.3 Saccharomyces cerevisiae as a Model Organism to Study Transcription
Introduction

tools is needed to read out quantitatively activity of single genes as a function of different activation levels, as well as mathematical tool to interpret experimental data. Genetic engineering and synthetic biology combined with mathematical modeling represent a potentially useful approach to decode the role of promoters in filtering signals from the binding of regulators to changes in gene expression. Moreover, the yeast *Saccharomyces cerevisiae* would represent a model organism of choice in providing support for this approach because particularly keen to studies on eukaryotic transcriptional regulation and suited to genetic engineering.

2.3 *Saccharomyces cerevisiae* as a Model Organism to Study Transcription

The budding yeast *Saccharomyces cerevisiae* is a unicellular eukaryote belonging to fungi of the phylum ascomycota and represents a powerful model organism in molecular biology especially to study gene regulation. Yeast is currently a lead organism in eukaryotic genomics for several reasons that makes it particularly favorable as model organism: Yeast can be easily cultured in laboratory and reproduces by a division process called budding in a generation time of approximately 90 min. Several commercially available strains contain mutations in metabolic genes that can be restored and used as selection markers for transformation. Yeast can in fact be transformed very easily by adding or deleting genes through homologous recombination and its capacity to undergo haploid or diploid life cycles offers opportunities for flexible genetic manipulation. Moreover, *Saccharomyces cerevisiae* is devoid of the large amount of non-coding DNA present in higher eukaryotes. Actually, haploid yeast genomes can be combined upon mating: Haploid cells of mating type A can undergo mating with cells of mating type α producing diploids that are equally stably cultured as the haploid ones. Yeast was the first eukaryote having been sequenced entirely in 1996 and since then databases such as the 'Saccharomyces Genome Database (SGD)' [55] are regularly curated and updated. The genome of budding yeast consists of approximately 6,000 genes which are packed in 16 chromosomes and represents a size of 12 Mbp. The knowledge related to yeast gene regulation can typically be transferred to more complex context of higher eukaryotes where the mechanism of transcription and its regulation are largely conserved.

2.4 The Transcription Machinery of *Saccharomyces cerevisiae*

The components of the yeast's transcription machinery are highly conserved from yeast to humans and thereby studies have been vastly based on *Saccharomyces cerevisiae* to understand basic mechanisms underlying eukaryotic transcription and gene expression.

Eukaryotic RNA polymerase II is an enzyme composed of 12 subunits (12 subunits in yeast and humans as well) that catalyzes the synthesis reaction of mRNA principally (it also synthesizes small nuclear RNAs (snRNA) and microRNAs). In addition, RNA polymerase I transcribes ribosomal RNAs (rRNAs) and RNA polymerase III transcribes small RNAs such as transfer RNAs (tRNAs).

General transcription factors (GTFs) or basal transcription factors are required for the initiation complex to be functional: TFIIB, TFIID, TFIIE, TFIIF, and TFIIH are basal transcription factors that assemble together with RNA polymerase II to form the preinitiation complex in order to initiate transcription of mRNA. In fact, differently

2.4 The Transcription Machinery of Saccharomyces cerevisiae — Introduction

from prokaryotes, RNA polymerase II cannot recognize target promoters directly: The preinitiation complex of prokaryotes requires the σ-factor and the unique prokaryotic RNA polymerase only.

On the other side, promoter DNA sequences, to which transcription factors bind, consist of core or basal promoter elements, promoter proximal elements, and distal enhancer elements. The core promoter is the minimal portion of the promoter required to properly initiate transcription. It is essentially constituted by a TATA element (located approximately 25 bp upstream of the transcription start) and a pyrimidine-rich initiator element (located at the start site). RNA polymerase II binds the core promoter and initiate transcription upon assembly of basal transcription factors (TFIID, TFIIB, TFIIF, TFIIE, and TFIIH) into a preinitiation complex. TFIID is a protein conserved among different organisms and is responsible for binding with the TATA box. TBP (TATA box binding protein), along with a variety of TBP-associated factors, constitute the TFIID. *In vitro*, TFIID can remain bound to its position in the core promoter after the polymerase has initiated transcription and so it can support transcription re-initiation. Promoter proximal elements are generally situated between 50 and 200 bp upstream of the start site and transcriptional regulators binding to these sequences can modulate transcription. Finally, distal enhancer elements, which can be found far from the transcription initiation site in either direction and orientation, constitute another group of DNA targets for factors modulating RNA polymerase II activity (see Fig.1).

Regulation of transcription in response to developmental or environmental signals is achieved by controlling assembly of the preinitiation complex or the catalytic efficiency of RNA polymerase II during initiation, elongation, or termination. Transcriptional activators can enhance transcription by increased recruitment and/or stabilization of the initiation complex on the promoter while transcriptional repressors prevent transcription by blocking the interaction between activators and RNA polymerase II complex. ([56]).

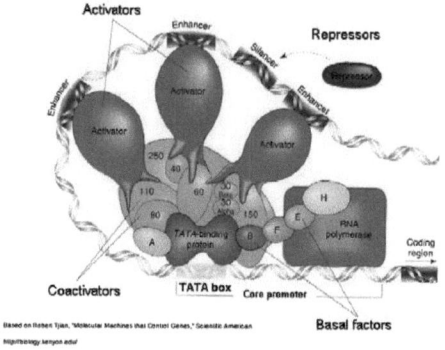

Figure 1: Schematic representation of transcription machinery organization. This example illustrates the disposition of different elements participating to higher eukaryote's transcriptional initiation.

2.5 Ligand Binding & Enzyme Kinetics

Historically, the study of enzyme kinetics and ligand binding contributed to a remarkable advance in the field of biochemistry by characterizing processes involving interactions between biomolecules [57, 58, 59, 60, 61]. Today the field is well established and routinely applied in the industrial development of chemical active compounds against enzymatic targets as well as in many other areas [62, 63, 64, 65, 66]. With the development of enzyme kinetics, properties and mechanisms of enzymes could be elucidated. Consequently, successful applications could take place in designing therapeutically active molecules to control diseases based on unrestrained activity of enzymes that escaped regulation [67, 68, 69]. Indeed, introducing the mathematical formalism of chemical kinetics theory to the study of biochemical reactions allowed to access important mechanistic insights in the dynamics of enzymes and other biological macromolecules, thereby improving significantly the understanding of biological processes at the molecular level. Physical properties of biochemical components could therefore be taken into consideration in the study of biological processes involving protein-protein interactions, interactions between proteins and nucleic acids, interactions of small organic molecules with biological macromolecules, enzyme-substrate interactions and transport processes [70, 71, 72]. Moreover, with this approach not only properties specific to given enzymes were found (e.g. specificity, stoichiometry and affinity), but also classes of mechanisms could be discovered (e.g. modes of inhibition, allosteric regulation, cooperative binding) which contributed to a significant improvement in the basic knowledge of general biological concepts.

Despite the encouraging aspects of approaching the study of biochemical processes mathematically, the framework of this formalism is almost completely based on *in vitro* conditions of experimentation. This implies to validate theoretical concepts with experimental conditions that are only partially consistent with the actual environment where the studied reactions occur. Actually, *in vitro* experiments assume conditions where the interacting molecular species are present as highly pure fractions and devoid of all other components normally present in their native context. Thereby, the development of theoretical tools can be biased by experimental validation that only partially represents the real process occurring *in vivo* [73, 74, 75, 76]. These problems can be due to the absence of molecular species contributing to the reaction (e.g. allosteric regulators), to the impact of the purification procedure on the macromolecules of interest, to the lack of heterogeneity characteristic of the intracellular environment, to the absence of the molecular crowding effect which implies different diffusion rates of macromolecules in cellular compartments with respect to simpler buffered chemical conditions of *in vitro* experiments. Moreover, the high level of connectivity in biological networks makes difficult to isolate pathways branches from the system they are part of.

Similarly to enzymatic reactions, several studies have been carried out in attempt to characterize the process of gene regulation at the level of transcriptional factors binding to DNA regulatory regions. However, like in enzyme kinetics, an important bottleneck relies on the incomplete picture that *in vitro* experiments give about the interaction of proteins with DNA. Moreover, in order to properly describe the real process of gene regulation occurring *in vivo* it is not sufficient to simply account for binding of regulators to DNA regulatory sequences but it is necessary to take into consideration the effect of several cellular components all together. Due to the large number of molecules participating to gene regulatory processes, such systems are obviously even more complicated to describe than enzymatic reactions: The process of gene regulation is composed of many steps and components that can be subject to regulation and thus can vary drastically under different conditions. Furthermore, if in the context of *in vitro* experimentation continuous deterministic models can be efficiently applied, *in vivo* description of gene regulation has shown to be subject to stochastic fluctuations [77, 78, 79]. The interaction of molecular partners like activators binding to DNA regulatory sequences studied *in vitro* can thereby differ very substantially from the one occurring *in vivo* [80, 33].

2.6 Mathematical Description of Gene Regulation

In order to improve our knowledge about the mechanisms underlying gene regulation, similarly to the advances that were determinant for the development of ligand binding and enzyme kinetics, there is the necessity of developing experimental techniques that allow to access biological parameters *in vivo* on one hand, and theoretical tools on the other hand to interpret the experimental findings.

Recently, besides of *in vitro* studies of protein-DNA interactions, a series of attempts to shed light on the process of gene activation *in vivo* have been proposed both on the theoretical and experimental level [15, 81, 82, 31, 25].

The practical consequence of dealing with living cells implies in fact to consider the effect of many components with the consequence that perturbations externally applied to study *in vivo* systems could affect undesired branches of the metabolism compromising cellular functions. Hence, since *in vivo* measurements of affinities are essential to understand the quantitative behavior of transcriptional regulatory networks in living cells [83], the current situation is clearly pointing out the necessity of developing new tools that can be applied to study living cells more efficiently. Importantly, new insights could be gained by applying minimally invasive experimental techniques that combined with the appropriate mathematical tools could provide maximal information. Our purpose is to adapt mathematical tools commonly used in enzyme kinetics to the context of gene regulation proposing an alternative modeling approach inspired by concepts widely used in different fields of science and engineering. These new mathematical tools combined with an experimental synthetic biology setup based on promoter engineering permit to access important information from experiments carried out in living cells.

2.7 Promoter Engineering to Study Gene Regulation

Nowadays, the understanding of the quantitative aspects of transcriptional regulation at the level of the promoter architecture is still poor and this is most notably due to ignorance of many biochemical parameters, especially their relevant *in vivo* values [82]. Moreover, many complex molecular events occur during regulated changes in transcription, and it is unclear to what extent changes in transcription factor concentrations can explain differences in expression between conditions [31].

A crucial point in studying transcriptional regulation is to be able to understand and predict how the architecture of promoters affects the transcriptional activity [1]. For the purpose of having a more complete understanding of the interactions involving the different molecular species that are determinant for gene regulation, we employed a technique based on libraries of engineered promoters to methodically investigate protein-DNA binding *in vivo*. It's in fact possible to apply modifications in gene regulatory sequences in order to study biomolecular interactions in the conditions where they naturally occur. Promoter libraries of variable strength for gene systems provide means to quantitatively describe the mechanisms driving regulation of gene expression. For example, applying series of graded alterations in a gene regulatory sequence that affect the binding affinity for a given molecular partner, would permit to modulate the strength of a given signal in order to extract information that characterizes the studied system. Moreover, a set of promoters of variable affinity can be obtained with techniques of directed evolution in order to constitute a library exhibiting broad ranges of genetic control [19, 21, 22].

Currently, despite the increasing amount of high-throughput experimental data and the consequent development of models to describe and relate them to their respective context, new experimental approaches are still needed to reveal important properties of biological systems in their native context [36]. The purpose of this methodology based on promoter engineering consists of the evaluation of microscopic parameters related to the binding events involved in gene regulation. In this way, parameters such as rate constants and cooperativity coefficients would become experimentally accessible in their native context.

2.8 Data Analysis & Mathematical Tools

The concept of *saturation function* is widely used in enzyme kinetics and ligand binding. Saturation functions are transfer functions that relate input and output in biochemical systems. They indicate the relationship between the concentration of substrate and rate in enzyme kinetics or the fraction of complex formation in molecular association studies as a function of the concentration of one system's component.

Saturation functions have been widely used to successfully interpret *in vitro* experiments of biochemical isolated systems. In addition to their application in enzyme kinetics and ligand binding, saturation functions can also be useful in biological systems in order to describe gene activation. Similarly to enzyme kinetics which studies how enzymes convert substrate into products, it can be assumed that DNA regulatory sequences exert a function comparable to enzymes. DNA regulatory sequences catalyze in fact RNA transcription which is proportional to gene expression so that degree of induction can be mapped to gene expression level. The tools applied to enzyme kinetics could thereby principally be employed in the context of gene regulation in order to describe gene regulatory processes.

2.8.1 Saturation Functions for Single Activator Binding

2.8.1.1 Rapid-Equilibrium Assumption to Derive Saturation Functions

The rapid-equilibrium or quasi-equilibrium approach is a simple way to derive saturations functions for chemical binding processes. The easiest way to model gene activation is to consider the activity of a DNA regulatory sequence (such as promoter sequences) to be proportional to the regulators bound to it. The variables to take into account are thereby the amount of regulator (A = activator) in the system and its propensity (K = binding affinity) to recognize the DNA sequence (B = receptor, binding site) forming a chemical complex (C = complex) in order to enhance transcription (D = product, RNA) upon recruitment of the transcriptional machinery. So that for a system composed of a single receptor binding site (considered here to be a promoter sequence with a single activator binding site) the saturation function can be derived assuming chemical equilibrium between receptor and activator (equilibrium assumption is supported by the fact

2.8 Data Analysis & Mathematical Tools

that association and dissociation reactions of activators to their binding sites are much faster than the process of transcription).

So considering the reaction

$$A + B \underset{k_{-1}}{\overset{k_1}{\rightleftharpoons}} C \xrightarrow{k_2} D \tag{1}$$

one is interested in expressing the complex formation C, which is proportional to the created product D, as a function of the reactants that can be kept under experimental control.

Hence, since reaction (1) assumes rapid equilibrium for the reactions connecting A, B and C compared to the reaction connecting C and D, i.e., $k_1, k_{-1} \gg k_2$, it holds that the association rate is

$$k_1[A][B] \tag{2}$$

and the dissociation rate

$$k_{-1}[C] \tag{3}$$

in which [] denote the concentrations of the relative species.

Assuming chemical equilibrium, association and dissociation rate balance each other so that we obtain

$$k_1[A][B] = k_{-1}[C] \tag{4}$$

and according to the low of mass action

$$K_A = \frac{[C]}{[A][B]} = \frac{k_1}{k_{-1}} \tag{5}$$

$$K_D = \frac{[A][B]}{[C]} = \frac{k_{-1}}{k_1} \tag{6}$$

Where K_A and K_D are the equilibrium association and dissociation constants, respectively.

Association and dissociation constants are functions of their standard states and can be reduced to the fundamental formula to

2.8 Data Analysis & Mathematical Tools — Introduction

$$\Delta G° = -RT \ln K° \qquad (7)$$

with $K°$ being the equilibrium constant and $\Delta G°$ the Gibbs free energy in the standard state, R the universal gas constant ($R = N_A \cdot k_B$ with N_A being the Avogadro constant and k_B the Boltzmann constant) and T the temperature.

Further, introducing the total concentrations of the different components participating to the reaction (which is the actual quantity that one can set experimentally) we obtain

$$[A]_t = [A] + [C] \qquad (8)$$

$$[B]_t = [B] + [C] \qquad (9)$$

In order to have C as a function of the other components of the system (C is supposed to be the chemical configuration that reflects activation of the system), and assuming that the activator A is in large excess with respect to its binding target B ($[A]_t \gg [B]_t$), it follows that $[A]_t \cong [A]$ and so we rearrange Eq.5 with Eq.9 to obtain

$$[C] = \frac{K_A[A][B]_t}{K_A[A]+1} = \frac{[A][B]_t}{[A]+K_D} \qquad (10)$$

Since the second step in reaction (1) is a first-order reaction, and that anyway due to the assumption of rapid-equilibrium between A and B, k_2 is the rate-determining step of the reaction, we can further work out the reaction rate for D

$$v = \frac{d[D]}{dt} = k_2[C] = \frac{k_2 K_A[A][B]_t}{K_A[A]+1} = \frac{k_2[A][B]_t}{[A]+K_D} \qquad (11)$$

Where $k_2[B]_t$ is the maximal possible level of system activation, in which every single DNA sequence has an activator bound to it ($[C] = [B]_t$). Consequently we call

$$V_{max} = k_2[B]_t \qquad (12)$$

and after substitution of Eq.12 in Eq.11 this results in the final form of the saturation function that reflects the fractional degree of system's activation:

2.8 Data Analysis & Mathematical Tools Introduction

$$v = \frac{d[D]}{dt} = \frac{V_{max}K_A[A]}{K_A[A]+1} = \frac{V_{max}[A]}{[A]+K_D} \qquad (13)$$

Assuming experimental conditions where the activator concentration can be considered as the system's variable and can be set externally, the graphical representation of the saturation function is a rectangular hyperbola called saturation or dose-response curve (see Fig.2).

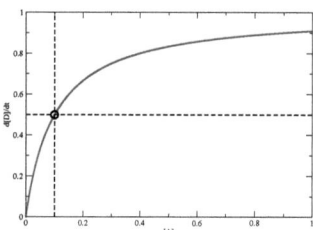

Figure 2: Graphical representation of Eq.13. The curve asymptotically tends to maximal activation of the system (here $V_{max} = 1$). The value of K_D on the x-axis corresponds to the y-axis value of $\frac{V_{max}}{2}$ highlighted by a circle of coordinates (K_D , $\frac{V_{max}}{2}$).

2.8.1.2 Steady-State Assumption to Derive Saturation Functions

An alternative way to model gene regulation is assuming steady-state of the intermediates instead of equilibrium [84]. In this case it is not necessary for the complex C to be in chemical equilibrium with A and B in order to transcribe the product D. As it will be evident later, this assumption is a more general theory with respect to the the one explained above and it includes the rapid-equilibrium treatment as a special case.

So, considering the reaction

$$A + B \underset{k_{-1}}{\overset{k_1}{\rightleftharpoons}} C \xrightarrow{k_2} D \qquad (14)$$

the production rate of D is

$$v = \frac{d[D]}{dt} = k_2[C] \qquad (15)$$

2.8 Data Analysis & Mathematical Tools

together with the condition of Eq.9 and assuming that after the initial phase, the complex concentration remains constant at steady-state, this results in

$$\frac{d[C]}{dt} = k_1([B]_t - [C])[A] - (k_{-1} + k_2)[C] = 0 \tag{16}$$

The steady-state treatment, like the rapid-equilibrium method, assumes that the free activator A is in large excess with respect to its binding target B ($[A]_t \gg [B]_t$). Thus, since $[A]_t \cong [A]$, we can rearrange Eq.16 in order to express C as a function of the activator concentration:

$$[C] = \frac{k_1[A][B]_t}{k_1[A] + (k_{-1} + k_2)} \tag{17}$$

which substituted in Eq.15 gives

$$v = \frac{d[D]}{dt} = k_2[C] = \frac{k_2 k_1 [A][B]_t}{k_1[A] + (k_{-1} + k_2)} = \frac{k_2[A][B]_t}{[A] + \frac{k_{-1} + k_2}{k_1}} \tag{18}$$

where the term $\frac{k_{-1}+k_2}{k_1}$ is called Michaelis constant (K_M):

$$K_M = \frac{k_{-1} + k_2}{k_1} \tag{19}$$

which reduces to K_D in the special case where the complex formation reaction is much faster than the product conversion, $k_2 \ll k_{-1}$, as it is the case in gene regulation (activator binding to DNA regulatory sequence is faster than transcription).

Further, the maximal activation of the system is reached when very high concentration of activator fully saturate all molecules of receptor DNA. In this case $[C] = [B]_t$ and the rate of product formation $\frac{d[D]}{dt}$ reaches asymptotically maximal value called V_{max}

$$V_{max} = k_2 [B]_t \tag{20}$$

Equation Eq.18 can then be rewritten in the form commonly known as *Michaelis-Menten equation*

2.8 Data Analysis & Mathematical Tools

Michaelis-Menten Equation:

$$v = \frac{d[D]}{dt} = \frac{V_{max}[A]}{[A] + K_M} \qquad (21)$$

The resulting dose-response curve obtained by this steady-state treatment is a rectangular hyperbola as for the rapid-equilibrium treatment for one binding site (see Fig.3). The only difference is the x-axis value corresponding to $\frac{V_{max}}{2}$ which is equals to $K_M = \frac{k_{-1}+k_2}{k_1}$ while using the rapid-equilibrium approach $\frac{V_{max}}{2}$ corresponds to a x-coordinate of $K_D = \frac{k_{-1}}{k_1}$.

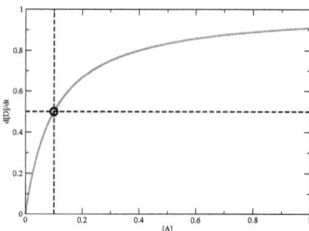

Figure 3: Graphical representation of Eq.21. The curve asymptotically tends to maximal activation of the system (here $V_{max} = 1$). The value of K_M on the x-axis corresponds to the y-axis value of $\frac{V_{max}}{2}$ highlighted by a circle of coordinates (K_M , $\frac{V_{max}}{2}$).

2.8.2 Adair Equation for Multiple Activators Binding

Compared to the rapid-equilibrium approach, steady-state treatment is more complicated and requires greater effort to derive, especially for large systems composed of many intermediates. The steady-state method is not particularly difficult *per se* as a method, its application simply may generate very large expressions that are not always easy to handle.

In contrast, assuming rapid-equilibrium one can easily face large systems and derive saturation functions that account for multiple ligands binding as well as for additional effects (e.g. binding cooperativity) in a straightforward way (see Fig.4).

The Adair equation is a rapid-equilibrium based method originally derived to describe the binding of oxygen to hemoglobin [24]. It can anyway be used efficiently to describe any binding process involving multiple ligand binding to receptors as well as cooperative binding.

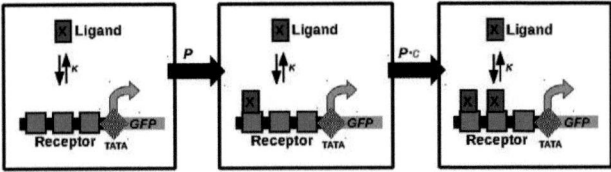

Figure 4: Sequential binding process described by Adair equation modeling approach. The receptor binds the first ligand with probability P, the probability of further ligand binding is influenced by a factor c determined by receptor-bound ligands. Ligands previously bound to the receptor can enhance further binding (positive cooperativity, $c > 1$) or hinder further binding (negative cooperativity, $c < 1$). If $c = 1$ there is no influence of bound ligands on subsequent binding events (independent binding).

2.8 Data Analysis & Mathematical Tools Introduction

2.8.2.1 General Adair Equation for Enzyme Kinetics

The Adair method consists of using the definition of chemical equilibrium binding constants and stoichiometric binding constants in order to formulate a general saturation function that can account for multiple ligand binding. The Adair equation allows a *mechanistic* definition of cooperativity that may be compared to empiric definitions such as the one represented by the Hill equation approach (discussed later).

The principle of the Adair equation is to sum all contributions participating to the activation of the described system in the numerator and to normalize this quantity by all possible binding states the system can assume in the denominator:

$$Y = \frac{Active}{Active + Inactive} = \frac{Configurations\,(Ligand\,Bound)}{Configurations\,(Total\,Receptor)} = \frac{[L]_{Bound}}{[R]_{Total}} \quad (22)$$

So that in order to describe the general case of a receptor R with n binding sites for ligands L, which upon binding forms a ligand-receptor complex RL_i with i ligands bound to it, Eq.22 becomes:

$$Y = \frac{\sum_{i=1}^{n} i[RL_i]}{\sum_{i=0}^{n}[RL_i]} = \frac{[RL] + 2[RL_2] + 3[RL_3] + \ldots + n[RL_n]}{[R] + [RL] + [RL_2] + [RL_3] + \ldots + [RL_n]} \quad (23)$$

where the numerical value that multiplies every term is a weight that accounts for the contribution of receptor's configurations with i ligands bound. Here it is assumed that these configurations contribute i times to system's activation with respect to a receptor with a single ligand bound. This assumption is appropriate to describe enzymatic reactions where i catalytic subunits of an enzyme convert i substrate molecules to i product molecules. This is exactly i times the corresponding contribution of a single catalytic subunit.

The expressions for the individual steps are represented by the stoichiometric binding constants κ_i for the ith step in series of n steps as follows:

$$
\begin{aligned}
&\text{R} + \text{L} \rightleftharpoons \text{RL}_1 \quad && \kappa_1 = \frac{[RL_1]}{[R][L]} && [RL_1] = \kappa_1[R][L] \\
&\text{RL}_1 + \text{L} \rightleftharpoons \text{RL}_2 \quad && \kappa_2 = \frac{[RL_2]}{[RL_1][L]} && [RL_2] = \kappa_2[RL_1][L] = \kappa_1\kappa_2[R][L]^2 \\
&\text{RL}_2 + \text{L} \rightleftharpoons \text{RL}_3 \quad && \kappa_3 = \frac{[RL_3]}{[RL_2][L]} && [RL_3] = \kappa_3[RL_2][L] = \kappa_1\kappa_2\kappa_3[R][L]^3 \\
& && \vdots && \quad (24) \\
&\text{RL}_{n-1} + \text{L} \rightleftharpoons \text{RL}_n \quad && \kappa_n = \frac{[RL_n]}{[RL_{n-1}][L]} && [RL_n] = \kappa_n[RL_{n-1}][L] = \kappa_1 \ldots \kappa_n[R][L]^n
\end{aligned}
$$

2.8 Data Analysis & Mathematical Tools   Introduction

Using Eq.24 to reformulate Eq.23 such that Y becomes a function of the ligand concentration $[L]$ instead of the complex $[RL_i]$ we obtain

$$Y([L]) = \frac{\kappa_1[R][L] + 2\kappa_1\kappa_2[R][L]^2 + 3\kappa_1\kappa_2\kappa_3[R][L]^3 + \ldots + n\kappa_1\ldots\kappa_n[R][L]^n}{[R] + \kappa_1[R][L] + \kappa_1\kappa_2[R][L]^2 + \kappa_1\kappa_2\kappa_3[R][L]^3 + \ldots + \kappa_1\ldots\kappa_n[R][L]^n}$$

$$= \frac{\kappa_1[L] + 2\kappa_1\kappa_2[L]^2 + 3\kappa_1\kappa_2\kappa_3[L]^3 + \ldots + n\kappa_1\ldots\kappa_n[L]^n}{1 + \kappa_1[L] + \kappa_1\kappa_2[L]^2 + \kappa_1\kappa_2\kappa_3[L]^3 + \ldots + \kappa_1\ldots\kappa_n[L]^n} \tag{25}$$

The stoichiometric binding constant κ_i can be expressed as a function of the equilibrium binding constant K for the ith step as

$$\kappa_i = \frac{n-i+1}{i} K \tag{26}$$

in which $\frac{n-i+1}{i}$ is a combinatorial term that accounts for the binding probabilities of association and dissociation of further ligands. It accounts for the probability of a ligand to bind the receptor divided by the the probability to unbind. For example, if two binding sites are present on a receptor, there are 2 combinations for the first ligand to bind and 1 to unbind, while there is only 1 combination left for the second ligand to bind and 2 to unbind.

Further, in order to account for binding cooperativity, a numerical factor is introduced in Eq.25 to take into consideration the effect of bound ligands on subsequent binding events. Thus, upon binding the receptor, the first ligand influences the second ligand binding by a factor c_1. Subsequently, the third binding event occurs under the influence of the previously bound ligands, i.e., $c_1 c_2$, and so on:

1th binding: $\kappa_1 = n \cdot K$

2nd binding: $\kappa_2 = \dfrac{n-1}{2} \cdot c_1 \cdot K$

3rd binding: $\kappa_3 = \dfrac{n-2}{3} \cdot c_1 \cdot c_2 \cdot K$

4th binding: $\kappa_4 = \dfrac{n-3}{4} \cdot c_1 \cdot c_2 \cdot c_3 \cdot K$

$\vdots$

n^{th} binding: $\kappa_n = \dfrac{1}{n} \cdot c_1 \cdot c_2 \cdot c_3 \ldots c_{n-1} \cdot K$

2.8 Data Analysis & Mathematical Tools — Introduction

If cooperativity terms are greater than unity, subsequent ligand binding is facilitated (positive cooperativity: $c_i > 1$), if smaller than unity the next binding event is hindered (negative cooperativity: $c_i < 1$) and independent binding occurs when the $c_i = 1$.

So, for the equilibrium binding configuration with i ligands bound (i.e., up to the ith step in the binding sequence composed of n steps) we can write

$$\kappa_1 \ldots \kappa_i = \sigma_i \gamma_i K^i \tag{27}$$

where

$$\sigma_i = \prod_{j=1}^{i} \frac{n-j+1}{j} \tag{28}$$

is the combinatorial term that accounts for the binding probabilities of association and dissociation of further ligands to the actual receptor configuration and

$$\gamma_i = \prod_{j=1}^{i} c_j^{i-j} \tag{29}$$

is the cooperativity term that considers the effect that ligands bound to the receptor exert on the next ligand binding.

The resulting form of the Adair equation is

Enzyme-Kinetics-Adair Equation:

$$Y([L]) = \frac{\gamma_1 \sigma_1 K[L] + 2\gamma_2 \sigma_2 K^2 [L]^2 + 3\gamma_3 \sigma_3 K^3 [L]^3 + \ldots + n\gamma_n \sigma_n K^n [L]^n}{1 + \gamma_1 \sigma_1 K[L] + \gamma_2 \sigma_2 K^2 [L]^2 + \gamma_3 \sigma_3 K^3 [L]^3 + \ldots + \gamma_n \sigma_n K^n [L]^n} \tag{30}$$

that can also be expressed in more compact form as

$$Y([L]) = \frac{\sum_{i=1}^{n} i \gamma_i \sigma_i K^i [L]^i}{1 + \sum_{i=1}^{n} \gamma_i \sigma_i K^i [L]^i} = \frac{\sum_{i=1}^{n} i (\prod_{j=1}^{i} c_j^{i-j})(\prod_{j=1}^{i} \frac{n-j+1}{j}) K^i [L]^i}{1 + \sum_{i=1}^{n} (\prod_{j=1}^{i} c_j^{i-j})(\prod_{j=1}^{i} \frac{n-j+1}{j}) K^i [L]^i} \tag{31}$$

which is represented graphically in Fig.5.

Due to the contribution weights for receptors bound by multiple ligands, the values assumed by the Adair equation range from 0 to the number of receptor binding sites n:

$$0 \leq Y([L]) \leq n \tag{32}$$

2.8 Data Analysis & Mathematical Tools

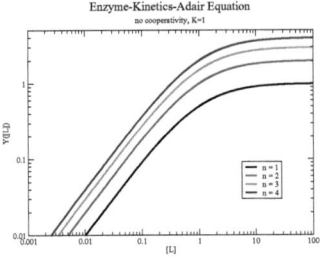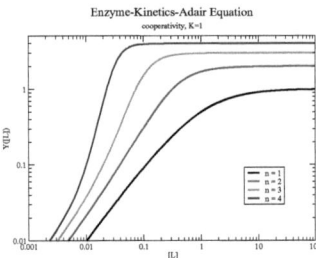

Figure 5: Enzyme-kinetics-Adair equation. The effect of a different number of ligand binding sites n (from 1 to 4) is compared in absence ($c_1, c_2, c_3 = 1$, $K = 1$, left panel) and presence ($c_1, c_2, c_3 = 10$, $K = 1$, right panel) of binding cooperativity.

2.8.2.1.1 Example: Hemoglobin Modeling with Adair Equation

Binding properties of the oxygen transport proteins hemoglobin and myoglobin were known before the end of the 19th century. Early studies of oxygen binding to these two similar proteins revealed that the saturation curve of myoglobin was a rectangular hyperbola while hemoglobin showed a sigmoidal curve. This allowed Adair to show that hemoglobin was composed of four subunits (actually now it is known that hemoglobin is a heterotetramer of type $\alpha_2\beta_2$) while myoglobin was a monomer.

The cooperative behavior responsible for the sigmoidal saturation curve of hemoglobin was modeled in a sequential way. The binding of the first oxygen molecule to one of the four sites enhances binding of the second molecule by a factor c_1. Subsequently, the binding of the second oxygen molecule enhances the third by a factor c_2 and finally, the third oxygen bound to hemoglobin promotes the last binding by a factor c_3.

Assuming the microscopic binding affinity of an oxygen molecule to one binding site being K (considered to be an equilibrium association constant), the sequential binding events are driven by the following stoichiometric affinities:

2.8 Data Analysis & Mathematical Tools   Introduction

1th binding: $\quad \kappa_1 = 4 \cdot K$

2nd binding: $\quad \kappa_2 = \dfrac{3}{2} \cdot c_1 \cdot K$

3rd binding: $\quad \kappa_3 = \dfrac{2}{3} \cdot c_1 \cdot c_2 \cdot K$

4th binding: $\quad \kappa_4 = \dfrac{1}{4} \cdot c_1 \cdot c_2 \cdot c_3 \cdot K$

where the numerical values $4, \frac{3}{2}, \frac{2}{3}, \frac{1}{4}$ are combinatorial factors accounting for the probability of an activator binding divided by the probability of an activator unbinding in relation to the free/occupied binding sites available (i.e., for the first binding there are 4 combinations for the ligand to bind and 1 to unbind, for the second binding there are 3 combinations left for the ligand to bind and 2 to unbind, and so on). The combinatorial factor for the ith ligand binding to a receptor of n binding sites is defined by $\frac{n-i+1}{i}$ which connects stoichiometric to binding constants (Eq.26).

As explained previously, the Adair approach consists of summing all configurations contributing to the activation of the system (in this case contributing to oxygen transport) normalized by all binding configurations the system can assume. The resulting expression is a saturation function that reflects the level of system activation as a function of its induction (in this case exposure to different oxygen concentrations, represented by x):

$$Y(x) = \frac{1 \cdot \sigma_1 \gamma_1 K x + 2 \cdot \sigma_2 \gamma_2 K^2 x^2 + 3 \cdot \sigma_3 \gamma_3 K^3 x^3 + 4 \cdot \sigma_4 \gamma_4 K^4 x^4}{1 + \sigma_1 \gamma_1 K x + \sigma_2 \gamma_2 K^2 x^2 + \sigma_3 \gamma_3 K^3 x^3 + \sigma_4 \gamma_4 K^4 x^4} \qquad (33)$$

In Eq.33, the numerical value that multiplies each term is the contribution weight that accounts for the relative contribution of every configuration to activation of the system: For the present example of oxygen transport, the term of the hemoglobin configuration with four oxygen molecules bound is multiplied by four because this molecule transports exactly four times the amount of oxygen than a molecule in the first configuration that transports one single oxygen molecule. Furthermore, the combinatorial factor for the ith ligand binding is defined by σ_i (Eq.28) and the cooperative binding factor by γ_i (Eq.29).

Combining weighted contributions and combinatorial factor σ_i Eq.33 becomes

$$Y(x) = \frac{1 \cdot 4 \cdot K x + 2 \cdot \frac{3}{2} \cdot 4 \cdot c_1 K^2 x^2 + 3 \cdot \frac{2}{3} \cdot \frac{3}{2} \cdot 4 \cdot c_1^2 c_2 K^3 x^3 + 4 \cdot \frac{1}{4} \cdot \frac{2}{3} \cdot \frac{3}{2} \cdot 4 \cdot c_1^3 c_2^2 c_3 K^4 x^4}{1 + 4 \cdot K x + \frac{3}{2} \cdot 4 \cdot c_1 K^2 x^2 + \frac{2}{3} \cdot \frac{3}{2} \cdot 4 \cdot c_1^2 c_2 K^3 x^3 + \frac{1}{4} \cdot \frac{2}{3} \cdot \frac{3}{2} \cdot 4 \cdot c_1^3 c_2^2 c_3 K^4 x^4}$$

$$= \frac{4 K x + 12 c_1 K^2 x^2 + 12 c_1^2 c_2 K^3 x^3 + 4 c_1^3 c_2^2 c_3 K^4 x^4}{1 + 4 K x + 6 c_1 K^2 x^2 + 4 c_1^2 c_2 K^3 x^3 + c_1^3 c_2^2 c_3 K^4 x^4} \qquad (34)$$

While the numerator accounts for all the molecular weighted configurations contributing to oxygen transport, the denominator accounts for all possible configurations the system can assume: In this case the unique configuration present in the denominator only is the one representing the empty receptor with four free binding sites. This configuration with zero ligands bound is calculated according to $\sigma_0 K^0 x^0 = 1$.

The Adair equation is thus a measure of activation of the hemoglobin system as a function of its induction (oxygen concentration). As a consequence of the weights for the different configurations in the numerator, Adair equation values range from 0 to the number of receptor binding sites ($0 \leq Y(x) \leq 4$). Saturation curves of oxygen transporter proteins hemoglobin and myoglobin with four and one binding sites, respectively, are represented graphically in Fig.6.

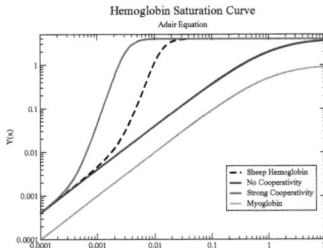

Figure 6: Comparison of different saturation curves that describe hemoglobin using the Adair equation (Eq.34). The dashed curve corresponds to values fitted to the sheep hemoglobin with parameters $a = 36$, $b = 1.68$, $c = 4.69$, $K = 1$. The blue curve represents absence of cooperativity ($a, b, c = 1$, $K = 1$) and the red curve strong cooperativity ($a = 10, b = 100, c = 1000$, $K = 1$). The green curve shows the corresponding situation for a monomeric protein like myoglobin.

2.8.2.2 Adaptation of the Adair Equation to Gene Regulation

The numerical weight for system's active configurations present in the numerator of the Adair equation described above is what makes the difference between the enzyme-kinetics-Adair equation explained in the previous section and the gene-regulation-Adair equation described in this section. For the reasons explained previously, modeling enzymes composed of multiple subunits implies to use a weight for binding configurations representing receptors bound with multiple ligands as a multiple of the single ligand bound configuration. In contrast, in order to model gene regulation from the perspective of DNA regulatory sequences containing multiple activator binding sites, one can assume that, independently on the number of activators bound, there is always the formation of only one initiation complex recruiting a RNA polymerase II. It has actually been shown that single activators bound to promoters can maximally activate gene ex-

2.8 Data Analysis & Mathematical Tools — Introduction

pression [77]. It is likely that the presence of multiple activators on DNA regulatory region contributes to gene activation mainly by increasing the recruitment probability of the RNA polymerase II. This represents a fundamental difference with respect to the enzyme-kinetics-Adair equation where multiple binding sites contribute to increase the maximum activation level of the system. In contrast, in the gene-regulation-Adair equation it is supposed that all binding configurations attain the same maximal level at saturation (compare Fig.5 with Fig.7).

Which means that the general principle

$$Y = \frac{Active}{Active + Inactive} = \frac{Configurations\,(Activator\,Bound)}{Configurations\,(Total\,Receptor)} = \frac{[A]_{Bound}}{[R]_{Total}} \qquad (35)$$

becomes for the general case of a receptor R with n binding sites for activators A:

$$Y = \frac{\sum_{i=1}^{n}[RA_i]}{\sum_{i=0}^{n}[RA_i]} = \frac{[RA] + [RA_2] + [RA_3] + \ldots + [RA_n]}{[R] + [RA] + [RA_2] + [RA_3] + \ldots + [RA_n]} \qquad (36)$$

Consequently, the numerical weight for the system's active configurations present in the numerator of the enzyme-kinetics-Adair equation is removed from Eq.30 and Eq.31 resulting in the gene-regulation-Adair equation as a function of the activator concentration $[A]$

Gene-regulation-Adair Equation:

$$Y([A]) = \frac{\gamma_1\sigma_1 K[A] + \gamma_2\sigma_2 K^2[A]^2 + \gamma_3\sigma_3 K^3[A]^3 + \ldots + \gamma_n\sigma_n K^n[A]^n}{1 + \gamma_1\sigma_1 K[A] + \gamma_2\sigma_2 K^2[A]^2 + \gamma_3\sigma_3 K^3[A]^3 + \ldots + \gamma_n\sigma_n K^n[A]^n} \qquad (37)$$

and in the more compact form

$$Y([A]) = \frac{\sum_{i=1}^{n} \gamma_i \sigma_i K^i [A]^i}{1 + \sum_{i=1}^{n} \gamma_i \sigma_i K^i [A]^i} = \frac{\sum_{i=1}^{n}(\prod_{j=1}^{i} c_j^{i-j})(\prod_{j=1}^{i} \frac{n-j+1}{j}) K^i [A]^i}{1 + \sum_{i=1}^{n}(\prod_{j=1}^{i} c_j^{i-j})(\prod_{j=1}^{i} \frac{n-j+1}{j}) K^i [A]^i} \qquad (38)$$

in which, the range of the values assumed by this form of the Adair equation is

$$0 \leq Y([A]) \leq 1 \qquad (39)$$

2.8 Data Analysis & Mathematical Tools

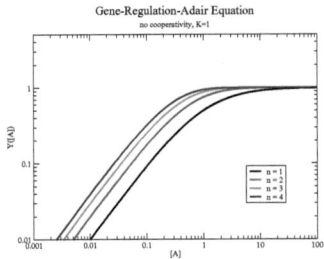 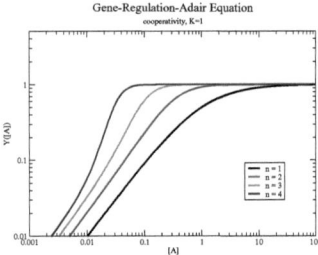

Figure 7: Gene-regulation-Adair equation to compare the effect of a different number of activator binding sites n (from 1 to 4) in absence ($c_1, c_2, c_3 = 1$, $K = 1$, left panel) and presence ($c_1, c_2, c_3 = 10$, $K = 1$, right panel) of binding cooperativity.

2.8.3 The Hill Equation

The Hill equation represents an empirical saturation function proposed by the British physiologist Hill [85] to describe the observed cooperative binding of oxygen to hemoglobin.

The model assumes that hemoglobin is present in a monomeric or multimeric configuration in absence or presence of oxygen, respectively. So that considering R (hemoglobin) and L (oxygen), according to Eq.24 we have for a sequential binding process with n steps

$$\kappa = \frac{[RL_n]}{[R][L]^n} \qquad (40)$$

where

$$\kappa = \kappa_1 \ldots \kappa_n \qquad (41)$$

and in order to express it as a saturation function of $[L]$ (similarly to Eq.25)

$$Y([L]) = \frac{[L]_{Bound}}{[R]_{Total}} = \frac{n[RL_n]}{[R] + [RL_n]} = \frac{n\kappa[R][L]^n}{[R] + \kappa[R][L]^n} = \frac{n\kappa[L]^n}{1 + \kappa[L]^n} \qquad (42)$$

According to Eq.27 and temporarily neglecting the sequential cooperative factors γ (the reason will be explain in the next paragraph) we obtain

$$\kappa = \sigma_n K^n = K^n \qquad (43)$$

2.8 Data Analysis & Mathematical Tools — Introduction

since in a n step binding process $\sigma_n = 1$ then κ can be substituted with K^n so that Eq.42 becomes

$$Y([L]) = \frac{nK^n[L]^n}{1 + K^n[L]^n} = \frac{n[L]^n}{K_D^n + [L]^n} \qquad (44)$$

Compared to the Adair equation for enzymatic reactions, Eq.44 only contains terms for empty and saturated receptors (1 and $[L]^n$, respectively). Thus, this procedure to derive the Hill equation assumes only two extreme states of the system: Fully off (empty receptor) and fully on (saturated receptor). Consequently, this implies *a priori* infinite binding cooperativity. Namely, the binding process described by the Hill equation means that upon first ligand binding, the empty receptor (inactive state) immediately switches to the fully bound configuration (maximally activated state) without intermediate states. Eq.44 describes indeed this idealized situation of infinite binding cooperativity. Anyway, in reality intermediate states of partially saturated receptors do exist and thus the exponent n will always be smaller then the number of receptor binding sites. So, the exponent is written n_H (called *Hill coefficient*) and Eq.44 becomes

$$Y([L]) = \frac{nK^{n_H}[L]^{n_H}}{1 + K^{n_H}[L]^{n_H}} = \frac{n[L]^{n_H}}{K_D^{n_H} + [L]^{n_H}} \qquad (45)$$

where

$$0 < n_H < n \qquad (46)$$

However, despite the fact that the Hill coefficient is devoid of physical meaning and should be considered as a purely empirical value, this approach is particularly useful because it allows to set a maximal threshold in the cooperativity level: With infinite sequential cooperativity between binding steps, the measured exponent is equal to the number of ligand binding sites n. This value is in reality never reached, anyway the higher the cooperativity of the system, the closer the measured exponent n_H will tend to this idealized upper bound n providing an efficient quantitative measure for binding cooperativity:

Positive cooperativity: $n_H > 1$
Independent binding: $n_H = 1$
Negative cooperativity: $n_H < 1$

Where positive/negative cooperativity results in facilitated/hindered further binding with respect to the actual receptor configuration, while independent binding implies that each binding event has no influence on the next (see Fig.8).

The Hill equation is commonly written similarly to the Eq.13 using maximal activation term V_{max}, which here anyway acts as a simple scaling factor, and since its wide range of application in different contexts, the independent variable is substituted with x that can account for different inputs (e.g. enzymatic substrate concentration, transcriptional activator concentration) resulting in

Hill Equation:

$$Y(x) = \frac{V_{max} \cdot x^{n_H}}{x^{n_H} + K_D^{n_H}} \qquad (47)$$

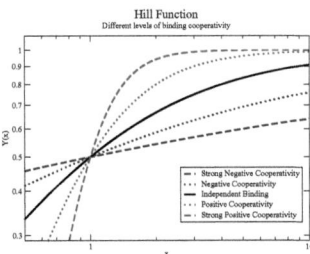

Figure 8: Hill function represented with different levels of binding cooperativity. Strong negative cooperativity ($n_H = 0.25$), negative cooperativity ($n_H = 0.5$), independent binding ($n_H = 1$), positive cooperativity ($n_H = 2$), strong positive cooperativity ($n_H = 4$); $V_{max} = K_D = 1$.

2.8.4 Logarithmic Sensitivity

Logarithmic sensitivity or logarithmic derivative or also called elasticity is a key tool that allows to correlate input to output in systems of different contexts and it is widely used in different fields outside of chemistry like finance and management, as well as in engineering and computer science. Logarithmic sensitivity expresses the relative increase of a variable ($y =$ output) with respect to another variable ($x =$ input).

Logarithmic Sensitivity:

$$\varepsilon_x^y = \frac{\%\ change\ in\ y}{\%\ change\ in\ x} = \frac{\partial [ln(y)]}{\partial [ln(x)]} = \frac{x}{y} \cdot \frac{\partial y}{\partial x} \qquad (48)$$

The logarithmic sensitivity in chemistry has been applied to metabolic control analysis in order to study regulatory mechanisms in cellular networks. In chemical kinetics elasticity is used to determine the order of reactions with respect to a particular reactant. That is, for mass action rate laws elasticity assumes a numerical value that corresponds to the kinetic order of the reaction with respect to a specific reactant: For the reaction

2.8 Data Analysis & Mathematical Tools — Introduction

$$A + 2B \xrightarrow{k} C \qquad (49)$$

the reaction rate is

$$v = \frac{d[C]}{dt} = k[A][B]^2 \qquad (50)$$

and the logarithmic sensitivity with respect to A is

$$\varepsilon^v_{[A]} = \frac{\partial [ln(v)]}{\partial [ln([A])]} = \frac{[A]}{v} \cdot \frac{\partial v}{\partial [A]} = \frac{[A]}{(k[A][B]^2)} \cdot \frac{\partial (k[A][B]^2)}{\partial [A]} = 1 \qquad (51)$$

while the logarithmic sensitivity with respect to B is

$$\varepsilon^v_{[B]} = \frac{\partial [ln(v)]}{\partial [ln([B])]} = \frac{[B]}{v} \cdot \frac{\partial v}{\partial [B]} = \frac{[B]}{(k[A][B]^2)} \cdot \frac{\partial (k[A][B]^2)}{\partial [B]} = 2 \qquad (52)$$

So that it results that reaction Eq.49 is of first order in A and second order in B as it can be seen by a graphical representation of $ln(v)$ vs $ln[A]$ and $ln(v)$ vs $ln[B]$, respectively (see Fig.9).

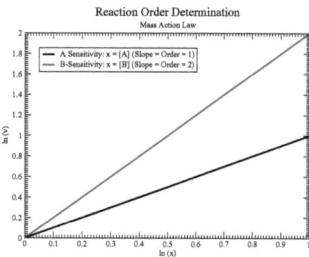

Figure 9: Kinetic order of the reaction determined by graphical representation of $ln(v)$ vs $ln[A]$ and $ln(v)$ vs $ln[B]$. The resulting slope corresponds to the respective order of the reaction which can be calculated using the logarithmic sensitivity with respect to A ($\varepsilon^v_{[A]}$) and B ($\varepsilon^v_{[B]}$) (equations Eq.51 and Eq.52, respectively).

2.8 Data Analysis & Mathematical Tools   Introduction

Alternatively, for more complex systems such as saturation functions the elasticity can also assume the form of a function of inputs. For example, calculating the logarithmic sensitivity of the Hill function (Eq.47) results in a function of the input x comprised between 0 and n_H (see Fig.10):

$$\varepsilon_x^{Y(x)} = \frac{n_H K_D^{n_H}}{K_D^{n_H} + x^{n_H}} \qquad (53)$$

$$0 \leq \varepsilon_x^{Y(x)} \leq n_H \qquad (54)$$

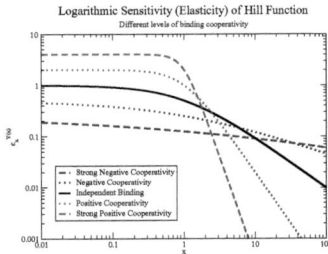

Figure 10: Logarithmic sensitivity of Hill functions represented in Fig.8 with different levels of binding cooperativity. Strong negative cooperativity ($n_H = 0.25$), negative cooperativity ($n_H = 0.5$), independent binding ($n_H = 1$), positive cooperativity ($n_H = 2$), strong positive cooperativity ($n_H = 4$); $V_{max} = K_D = 1$.

The same calculation has been applied to the Adair equation for both enzyme kinetics and gene regulation as presented in the previous sections to highlight system's properties in case of multiple ligand binding combined with the effect of binding cooperativity (see Fig.11 and compare with Fig.5 and Fig.7).

2.8.5 Parametric Representation of Gene Expression Data

Parametric representation can be used to connect outputs of different systems as a function of a common input: When comparing systems that exhibit different responses to same inducing conditions, it can be convenient to express deviations between systems rather than their corresponding saturation functions. For example, parametric representation can be used to represent the response of distinct promoters reacting differently to a common stimulus (e.g. external induction) possibly evoking stronger response in certain genes with respect to others. A concrete example is found in the galactose network of *Saccharomyces cerevisiae* where induction is mediated by common regulatory components responsible for the activation of a set of different genes (GAL genes) upon

2.8 Data Analysis & Mathematical Tools   Introduction

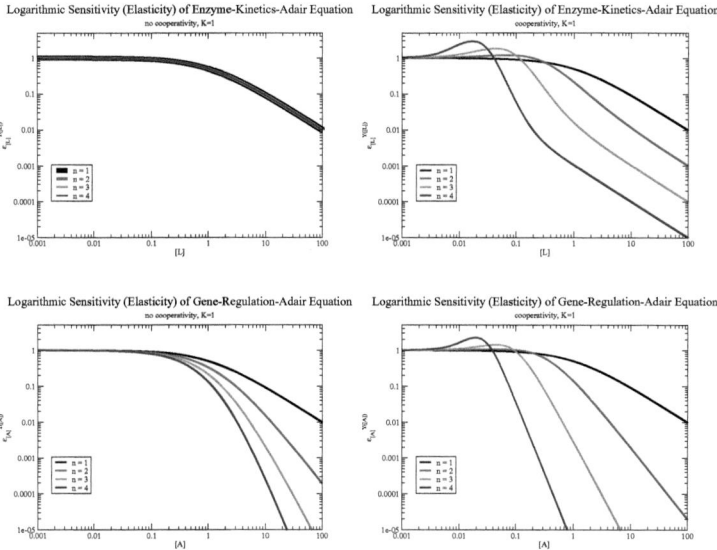

Figure 11: Logarithmic sensitivity of the enzyme-kinetics-Adair (upper panels, compare with Fig.5) and gene-regulation-Adair (lower panels, compare with Fig.7) equations to compare the effect of a different number of activator binding sites n (from 1 to 4) in absence ($c_1, c_2, c_3 = 1$, $K = 1$, left panels) and presence ($c_1, c_2, c_3 = 10$, $K = 1$, right panels) of binding cooperativity.

galactose exposure: The same stimulus activates several genes each driving different responses.

Comparing *in vivo* measurements of gene expression data can result difficult due to experimental errors and this way of representing measurements data can be convenient to compensate external perturbations: If external noise affects a certain induction condition, the effect would propagate to all systems under correlation and the resulting representation would conserve its characteristic features.

A simple example to illustrate parametric representation consists of using the Adair equation to draw saturation curves of single and multiple binding sites systems exposed to a common input. Further, one combines the curves by expressing single binding site promoters on the x-axis and multiple binding site promoters on the y-axis. This representations is particularly useful for the purpose of determining the gain in expression due to multiple binding sites. The advantage consists first of all of a simple yet efficient reference system that can be used unambiguously as a fixed point in order to compare enrichment in gene activity correlated to multiple binding sites. The reference system consists of a diagonal identity function that represents the parametric combination of single operator (on the y-axis) *versus* single operator system (on the x-axis). This simply accounts for zero enrichment when comparing two systems (obviously since two equal systems are compared). More interestingly, in the same fashion as explained before but now comparing multiple operator (on the y-axis) *versus* single operator system (on the x-axis), the curve assumes the shape of a hyperbolic function and depending on the applied stimulus common to the analyzed systems, different portions of the curves become accessible, conserving anyway the same topology (see Fig.12). The expression gain between single and multiple operator systems can be quantified by the area between the two curves and information about the switch-like character of the system as a consequence of multiple binding sites is encoded in their slopes.

2.8.6 Logistic Function

Logistic function (also called the Verhulst model or logistic growth curve) is a standard sigmoidal function initially developed to describe population growth but frequently used to describe general natural processes (e.g. chemical autocatalytic reactions) which found applications in a range of different fields from biology to economics and engineering.

The differential equation describing the continuous logistic model is:

$$\frac{dN}{dt} = \frac{rN(K-N)}{K} \tag{55}$$

where r is the maximal population growth and K the carrying capacity that represents population's limit. Nondimensionalization of Eq.55 leads to

$$\frac{dx}{dt} = rx(1-x) \tag{56}$$

2.8 Data Analysis & Mathematical Tools — Introduction

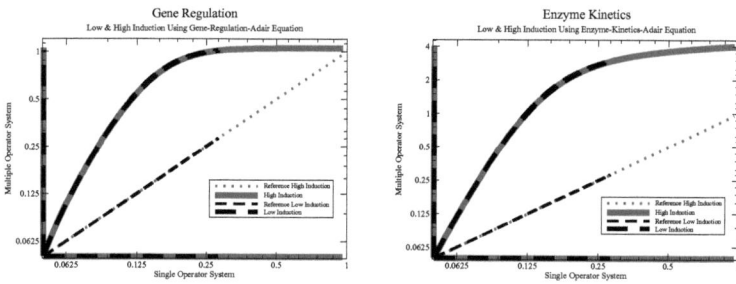

Figure 12: Saturation curves in parametric representation. Multiple operator curves are generated using the Adair equations (right plot: Enzyme-kinetics-Adair equation (Eq.31); left plot: Gene-regulation-Adair equation (Eq.38)) with $n = 4$, $K = 1$, $c_1 = 10$. $c_2 = c_3 = 1$. In addition, basal expression was set at 0.05 units and added to the respective equations. Induction was arbitrarily set at 10 and 0.3 corresponding to the denomination high and low induction, respectively. On the x-axis it is represented the range of expression covered by the single operator system while on y-axis by the multiple operator system. Curves corresponding to high (red lines) and low (black lines) induction are obtained by combining expression values of x and y-coordinates while reference systems are constructed by the same procedure but comparing single *versus* single operator system, thus giving a diagonal line (thin dotted and dashed lines).

in which $x = \dfrac{N}{K}$ and has the analytical solution

$$x(t) = (1 + (\frac{1}{x_0} - 1)e^{-rt})^{-1} \qquad (57)$$

In the context of our study we use the version of the logistic equation commonly used in the field of neural networks:

$$y = \frac{1}{1 + e^{-x}} \qquad (58)$$

which, with the addition of variable parameters can be transformed into a four-parameter logistic equation and efficiently used to carry out fitting procedures of sigmoidal curves (see Fig.13).

2.8 Data Analysis & Mathematical Tools

Four-Parameter Logistic Equation:

$$y = \frac{A0}{A1 + e^{-A2 \cdot x}} + A3 \qquad (59)$$

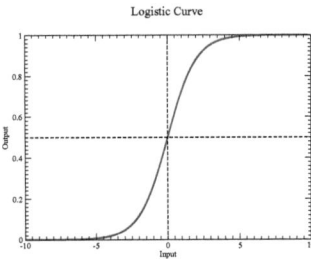

Figure 13: Standard logistic sigmoidal function (Eq.59) with $A0 = A1 = A2 = 1; A3 = 0$. $x =$ Input; $y =$ Output.

3 Material and Methods

3.1 Methods

The methods presented in this section concern all parts of the work detailed in the current manuscript.

3.1.1 Data Analysis

3.1.1.1 Flow Cytometry

Flow cytometry is a technology that allows to measure fluorescence intensity of single cells passing in a fluid stream through a laser beam. The laser excites specific fluorophores such as fluorescent proteins and a detector reads the longer wavelength fluorophore's emitted light. The outcoming signal is further converted into an electric signal and conveyed to a computer that displays different features of the sample properties. Flow cytometry was used in this study to measure the fluorescence level of yeast cells. This allows one to quantify reporter gene expression under different conditions which cells are exposed to with the advantage with respect to fluorescence microscopy that it permits to obtain information about a cell population simultaneously.

For the present study, cells were grown overnight in an appropriate selection medium at 30 °C. Subsequently, cells were transferred in induction tubes with same selection medium used for the overnight culture together with different concentrations of inducer (e.g. doxycycline). The initial cell concentration in induction tubes depends on the total induction time the cells are exposed to and the growth rate in the applied conditions. It has been counted that the final cell density after induction growth should be of $OD_{600nm} \approx 0.8$. After the period of induction, samples have been kept on ice and loaded into the flow cytometry instrumentation according to the instrument's user protocol. Mean cell fluorescence has been obtained from 5-15% of the total cell population selected from a gated region in the forward-scatter *versus* side-scatter plot. Mean fluorescence of 10,000-30,000 cells has been measured per induction condition. In addition to the induced samples, a non-induced control strain has been used to account for background fluorescence. The effective fluorescence has been then calculated according to the following formula:

$$Fluorescence = \frac{F}{C} - 1 \qquad (60)$$

where F corresponds to the fluorescence measure of the induced sample and C to the fluorescence of the non-induced control strain.

3.1.1.2 β-Galactosidase CPRG assay

LacZ assays have been performed following standard protocol of clontech [86]. LacZ is a commonly used reporter gene in transfection experiments because the gene product, β-galactosidase, is very stable and resistant to proteolytic degradation and easily assayed. The levels of active β-galactosidase expression can be quickly measured by its catalytic hydrolysis of CPRG (Chlorophenol red-β-D-galactopyranoside) substrate to a dark red product. The high sensitivity improves the measurement of β-galactosidase activity when the reporter gene expression is low.

3.1.1.3 Real Time PCR

Real time PCR is a variant of the classical polymerase chain reaction (PCR) but with the advantage that it permits to monitor the time evolution of the reaction and thereby to quantify the initial amount of DNA present in the sample. In the context of the present study, the real time PCR has been combined with reverse transcription in order to quantify messenger RNA of yeast cells. For this purpose messenger RNA has been firstly extracted from yeast's cell cultures and converted into cDNA through reverse transcription in order to be detected and quantified using real time PCR according to the instrument's user protocol.

3.1.2 Molecular Biology Methods

3.1.2.1 Yeast Transformation (Lithium Acetate Method)

Yeast transformation has been performed using the lithium acetate method according to the standard protocol described in [87]. The lithium acetate method is based on the fact that alkali cations make yeast competent to take up DNA. After yeast is briefly incubated in buffered lithium acetate, transforming DNA is introduced with carrier DNA. Addition of polyethylene glycol (PEG) and a heat shock trigger DNA uptake. The yeast have been then plated on selective media [88].

3.1.2.2 Bacteria Growth

E. coli cultures were grown at 37 °C in LB medium supplied with 75 µg/ml ampicillin.

3.1.2.3 Yeast Growth

Yeast cultures were grown at 30 °C in selection medium supplemented with 2% (w/v) carbohydrate source and inducer for a variable time (usually 6 hrs if not otherwise specified). In case of use of raffinose as carbohydrate source, the medium has been further supplemented with 0.005% of glucose in order to facilitate cell growth.

3.1.2.4 Molecular Cloning

pBluescript-based pRS vectors have been used to build synthetic constructs in the present work. These vectors contain a bacterial origin of replication (ORI), a bacterial selection marker (e.g. ampicillin), a yeast selection marker (e.g. URA3) and a multiple cloning site. Two types of pRS are available: Yeast integrative plasmids (YIp) which integrate into a particular locus in the yeast genome by homologous recombination and yeast centromeric plasmids (YCp) which contain a centromeric sequence (CEN) and a yeast's origin of replication (ARS). Different constructs have been build in this study using pRS vectors as starting material. Vectors were purified from *E. coli* cultures according to the plasmid isolation procedure described in the material section. The plasmids were subsequently engineered using modern molecular cloning techniques [89] and later re-transformed in *E. coli* for amplification. Constructs and intermediates between cloning steps were systematically sequenced to check for correctness before transformation in yeast cells.

3.1.2.5 Promoter Libraries

Promoter libraries used in this study were build by cloning synthetic binding sites specific for a transcription factor of interest in pRS306 vectors. pRS306 vectors have been used to construct the following libraries by insertion of the modified promoters and reporter gene blocks flanked by transcriptional terminators between *KpnI* and *NotI* restriction sites (see Appendix I for sequence details).

3.1.2.5.1 *tet*-Operator Library in *EGT2* Core Promoter

For the *tet*-operator library in P_{EGT2}, the binding sites for rtTA (*tet*-operators) have been synthesized as single stranded DNA first, annealed in a water bad gradient (with initial temperature of 100 °C and progressively cooled down to room temperature), phosphorylated, and subsequently cloned in the core promoter region of the yeast's *EGT2* gene upstream of a reporter gene. The multiple operator library has been constructed using restriction site sequences (*NheI, SpeI, SgsI, AvrII, XbaI*) as spacers between operators. The single operator library has been constructed using *AvrII* as restriction site for insertion. Constructs containing one to five operators were constructed in the same fashion where blocks of variable number of *tet*-operators have been inserted starting from the most proximal restriction site except for the single operator construct that has been constructed using *AvrII* as restriction site for insertion. *tetO2inphase* has been made by cutting in the restriction site between the two operators of *tetO2inEGT2* construct and DNA flapping regions klenow filled followed by blunt end self ligation thereby leading to a spacer sequence increase from 6 to 10 bp.

3.1.2.5.2 Gal4pBS Library in *GAL1* Core Promoter

For the Gal4pBS library in P_{GAL1}, the synthetic binding site sequences for the Gal4p activator were introduced in the *GAL1* promoter through fusion PCR. The modified promoters have been subsequently cloned into pRS306 vectors to obtain promoters and reporter gene blocks flanked by transcriptional terminators comparably to the 'tet-Operator Library in *EGT2* core promoter'.

3.1.2.6 Activators

3.1.2.6.1 rtTA

Fusion protein consisting of rtetR DNA binding domain, VP16 acidic transcriptional activation domain. The binding of rtTA to *tet*-operators can be adjusted through the applied doxycycline concentration (proportional to the active rtTA) [90].

In nature, the P_{tet} promoter expresses TetR, the repressor, and TetA, the protein that pumps tetracycline antibiotic out of the cell to confer resistance to gram-negative bacteria [91]. Expression of the tetracycline-resistance gene *tetA* as well as the tetracycline repressor gene *tetR* itself are subject to negative control by Tet repressor [92, 93, 94, 23]. VP16 is the transcriptional activation domain taken from herpes simplex virus.

3.1.2.6.2 GEV

Fusion protein consisting of the Gal4p DNA binding domain, an estradiol receptor domain and the VP16 transcriptional activation domain. The activity of GEV can be modulated by the applied estradiol concentration (proportional to the active GEV) [95].

3.1.2.6.3 tetR-Ssn6

Fusion protein consisting of tetR DNA binding domain and Ssn6 repressor domain. In absence of doxycycline, tetR-Ssn6 binds its target regulatory sequence and represses expression of the adjacent gene [96].

3.1.2.6.4 rtetR-Ace2p

Fusion protein consisting of rtetR DNA binding domain and the entire Ace2p protein. BswI and SgsI restriction sequences have been used as spacer sequence between the rtetR DNA binding domain and the Ace2p. The spacer (CGTACGGGGCGCGCC) is translated to ArgThrGlyArgAla connecting the two protein modules.

3.1.2.6.5 rtetR-Swi5p

Fusion protein consisting of rtetR DNA binding domain and the entire Swi5p protein. BswI and SgsI restriction sequences have been used as spacer sequence between the rtetR DNA binding domain and the Swi5p. The spacer (CGTACGGGGCGCGCC) is translated to ArgThrGlyArgAla connecting the two protein modules.

3.1.2.6.6 rtetR-NLS-Swi5pABCD

Fusion protein consisting of rtetR DNA binding domain and the ABCD domains of Swi5p protein according to the denomination used in [97, 98]. The two modules have been linked between each other by the nuclear localization sequence (NLS, region F according to the denomination used in [97, 98]). rtetR DNA binding domain (DNA sequence: nucleotides 1-643) has been connected to SWI5 nuclear localization sequence (DNA sequence: nucleotides 1900-2125) through a link consisting of GGGCGCGCC (translated in GlyArgAla), and the latter SWI5 sequence connected to SWI5 ABCD domains (DNA sequence: nucleotides 4-1639) through a link consisting of CCTGCAGGG (translated in ProAlaGly).

3.2 Materials

3.2.1 Data Analysis

3.2.1.1 Flow Cytometry

Flow cytometry experiments were performed using a Beckmann Coulter CYTOMICS FC 500 flow cytometer. The data where acquired using the CXP software.

3.2.1.2 β-Galactosidase CPRG assay

Buffer 1: 2.38 g HEPES, 0.9 g NaCl, 0.065 g L-Aspartate hemi Mg-Salt (Sigma), 1 g BSA, 50 µl Tween 20. Fill up to 100 ml. pH 7.25 - 7.3. Stored in the dark at 4 °C.

Buffer 2: 27.1 mg of CPRG in 20 ml of Buffer 1. This buffer should be prepared freshly before the experiments. It can be stored at 4 °C in the dark for 2-3 weeks.

Zinc Chloride Solution: 100 ml of 3 mM $ZnCl_2$ is prepared in water.

3.2.1.3 Real Time PCR

Real time PCR experiments have been performed using ABI® PRISM 7900HT Sequence Detection System (Applied Biosystems). An RNA extraction kit (Ambion) was used to purify RNA and reverse transcription was carried out with 'QuantiTect® Reverse Transcription Kit' (Quiagen) following the respective product protocols. The cDNA amplification is monitored using KAPA SYBR® FAST qPCR Kit (Kapabiosystems).

3.2.2 Molecular Biology Methods

3.2.2.1 Yeast Transformation

Tris-HCl: 10 mM Tris-HCl pH 7.5.

LiT Solution: Lithium acetate Tris solution: 100 mM lithium acetate in 10 mM Tris-HCl, pH 7.5.

Carrier DNA: Salmon sperm DNA (Sigma) denatured at 95 °C before use.

PEG: Polyethylene glycol (PEG) 4000 or 3350 dissolved in 100 mM LiT solution (1g:1ml).

YPAD: 1% yeast extract, 2% peptone, 2% glucose, 30 mg/liter adenine sulphate.

Selection Plates: 0.69% yeast nitrogen base (Foremedium), 2% glucose, 2% agar, 100 ml of 10X-concentrated solution of amino acid drop-out (Foremedium) in 1 L medium.

3.2.2.2 Plasmid Isolation from *E. coli* cultures

GenElute^{TM} Plasmid Miniprep Kit has been used according to the procedure described in product protocol.

3.2.2.3 Inducer Stock Solutions

Doxycycline: 5 M doxycycline stock solution is made in 50% ethanol. From this solution further diluted stock solutions can be made and all of them can be stored at -20 °C.

Estradiol: 5 M estradiol solution is made in 99% ethanol. From this solution an estradiol solution of 200 µM is diluted in DMSO. From this solution a 5 µM stock solution is made freshly before the experiment. 5 M and 200 µM solutions can be stored at -20 °C. Adding 1 µl of the 5 µM stock solution in 1 ml of medium makes the final estradiol concentration to 1 nM.

Galactose: 20% w/v galactose stock solution is made in water and then used to obtain final galactose concentrations in the media.

3.2.2.4 Bacteria Strains

SURE cells (Stratagene) were used for plasmid amplification and bacterial transformation.

3.2.2.5 Yeast Strains

All the strains used in this study were derived from the following EUROSCARF's strains:

S288C-Derived Strains

- BY4741 (MATa; his3Δ 1; leu2Δ 0; met15Δ 0; ura3Δ 0)
- BY4742 (MATα; his3Δ 1; leu2Δ 0; lys2Δ 0; ura3Δ 0)
- SY991 (MATa; ura3Δ 0; his3Δ 1; leu2Δ 0; trp1Δ 63; ade2Δ 0; lys2Δ 0; ADE8)
- SY992 (MATα; ura3Δ 0; his3Δ 1; leu2Δ 0; trp1Δ 63; ade2Δ 0; lys2Δ 0; ADE8)

W303 Wild Type Strains

- BMA64-1A (MATa; ura3-52; trp1Δ 2; leu2-3,112; his3-11; ade2-1; can1-100)
- BMA64-1B (MATα; ura3-52; trp1Δ 2; leu2-3,112; his3-11; ade2-1; can1-100)

4 Results

The results described in this thesis were obtained through different sets of experiments in which several genes and regulators or different variants of certain cellular components were used to address specific questions concerning gene regulation. The general character of the subject outlines the usefulness of brief introductory parts describing the experimental setup as well as summarizing paragraphs at the beginning of each section. This additional material aims to facilitate understanding and bridging between the different experimental findings.

4.1 Context-Dependent Effects on Gene Regulation

> *Motivation:* The role of gene specific effects concerning activator binding cooperativity is unknown. The context in which the process of gene regulation occurs is represented by the target promoter sequence on one hand, and on the other hand by activator's transcriptional activation domains.
> *Aims:* Understanding the influence of context specific effects on the process of gene activation in relation to multiple activator binding sites and activator binding cooperativity.
> *Results:* Gene specific effects play an important role in decoding induction stimuli. Promoter sequence, number of activator binding sites present in it and the activator type used to mediate induction influence gene response. The context determines the role of multiple binding sites and their influence on binding cooperativity.

4.1.1 Introduction

In attempt to gain insights in the process of gene regulation in a quantitative fashion, we systematically investigated how different transcription factor mediated inputs are integrated by cell cycle promoters and transduced into gene expression response. Using a synthetic experimental setup we therefore replaced native activator binding sites by *tet*-operators in the yeast promoters of the *EGT2* gene (P_{EGT2}, containing 6 native binding sites for Swi5p/Ace2p) and of the *SIC1* gene (P_{SIC1}, containing 2 native binding sites for Swi5p/Ace2p) in order to study the mode of activation driven by the different promoter sequences under same induction conditions. More precisely, we tried to decode the role played by different promoters in integrating the signal mediated by rtetR-based fusion proteins binding to *tet*-operators. These fusion proteins consist of an externally controllable DNA binding domain and a transcriptional activation domain derived from the native activators (Swi5p and Ace2p). In addition, the viral VP16 activation domain of rtTA was also used in the same experimental conditions.

The binding of rtetR DNA binding domain to *tet*-operators can be controlled externally through doxycycline concentration and so we could compare activation of the different promoters on one hand, and influence of the number of activator binding sites combined with the role of the different transcriptional activation domains on the other hand. With this approach binding cooperativity could be measured quantitatively by non-linear regression fitting and the nature of the observed cooperativity interpreted in relation to specific effects promoted by the different context. One could in fact determine if cooperativity is due to effects caused by the recruitment of specific components by a certain transcriptional activation domain (e.g. chromatin remodeling complexes), or alternatively, activation domain independent cooperativity would indicate stabilizing interactions between activators (e.g. protein-protein based, DNA induced fit).

4.1.2 System Description

4.1.2.1 Transcriptional Activators

Swi5p and Ace2p are cell cycle regulated transcriptional activators that recognize the same DNA binding sequences *in vitro* despite having different roles *in vivo*. Swi5p activates transcription of *HO* gene which encodes for the specific endonuclease needed for mating type switching. Ace2p activates transcription of *CTS1* gene which encodes for chitinase needed to degrade the cell wall between mother and daughter cells in the final stage of cytokinesis. Both *HO* and *CTS1* are expressed in the late G1 phase and *SWI5* and *ACE2* are transcribed in late G2 phase. In addition, Swi5p and Ace2p also activate other genes: *ASH1, CDC6, EGT2, RME1, SIC1, PCL2, PCL9* that play different roles in the early G1 phase of the cell cycle (see Fig.14).

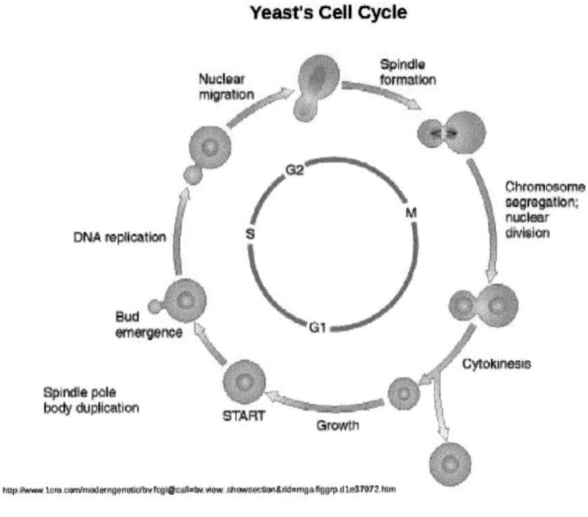

Figure 14: Representation of the cell cycle of *Saccharomyces cerevisiae* [99].

Swi5p and Ace2p are similar at the amino acid level and almost identical in the zinc finger DNA binding domain. Between Swi5p and Ace2p there is 83% amino acid identity and 94% similarity. The nuclear localization sequence of Swi5p and Ace2p is conserved and thus the nuclear localization during cell cycle is likely to be equivalent.

4.1 Context-Dependent Effects on Gene Regulation   Results

While some of the target genes mentioned above can be activated either by Swi5p or Ace2p, Swi5p seems to play the major role in activating other of these genes. The only marked difference is observed for the genes *HO* and *CTS1* where Swi5p and Ace2p are involved in specific activation. This effect may rely on the specific interaction that Swi5p undergoes with a specific molecular partner (Pho2p) binding to its activation domain. The main differences between Swi5p and Ace2p are in effect found in the activation domain region of the protein hinting for a potential specificity difference. [97].

Actually, another important difference between Swi5p and Ace2p is that only Swi5p remains active during mitosis. This allowed to reveal one more difference in relation to the role of the activation domain and indeed that Swi5p has the ability to recruit chromatin remodeling complexes (SWI/SNF complex) through its activation domain (region D based on the notation used in [97, 98]) (see Fig.15), thereby suggesting the ability of Swi5p to activate mitosis specific genes in a cell stage where DNA is highly structured into chromatin [98].

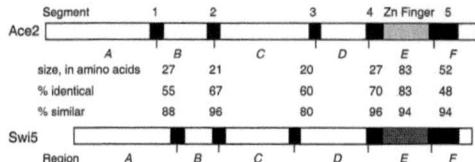

Figure 15: Scheme of the domains of Swi5p and Ace2p compared. Region ABCD is responsible for promoter specific activation. Region EF contains DNA binding domain and nuclear localization sequence [97].

Differently, Ace2p is not associated with such chromatin remodeling complexes and is active during interphase where the DNA doesn't require to be decompressed in order to allow gene activation. Due to these particular differences between two proteins highly conserved within the same organism, studying Ace2p and Swi5p represents an opportunity to investigate the role of different activation domains, and the consequent effects on gene activation. We therefore built genetic constructs in order to obtain fusion proteins composed of an externally controllable DNA binding domain and activation domains responsible for specific *in vivo* effects (see section 'Material and Methods'). The expression of fusion proteins was driven by the constitutive cycline 3 promoter (P_{CLN3}) and doxycycline was used to control their binding affinity.

4.1.2.2 Promoters

SIC1 and *EGT2* are cell cycle regulated genes containing multiple binding sites for the activators Swi5p and Ace2p. *SIC1* is one of the genes activated in mitosis and drives the expression of cycline dependent kinase (CDK) inhibitor that contributes to inactivate CDK/B cycline kinase and controls G1/S phase transition thereby preventing premature S phase entry and ensuring genomic integrity. Its promoter (P_{SIC1}) contains

2 binding sites for the activators Swi5p/Ace2p. *EGT2* is another gene expressed during the cell cycle and codes for an endoglucanase involved in cell separation in the G1 stage of the cell cycle. It is regulated in a cell cycle dependent manner. *EGT2* expression is highest between mitosis and early G1 [97, 98, 100]. The promoter of *EGT2* (P_{EGT2}) contains 6 binding sites for the activators Swi5p/Ace2p. The synthetic P_{EGT2} constructed for our purposes contains *tet*-operators substitutions for 5 and 6 of the 6 native binding sites ($P_{tetO5inEGT2}$ (the 3rd binding site is left unchanged) and $P_{tetO6inEGT2}$, respectively) while the synthetic P_{SIC1} has been substituted in both of its native binding sites ($P_{tetO2inSIC1}$). The activation curves of $P_{tetO5inEGT2}$ are informative to shed light on the effect of one native activator allowed to bind the regulatory sequence independently on external induction. This activator could previously modify the structure of the promoter (e.g. by chromatin remodeling) so that the externally controlled activators would bind $P_{tetO5inEGT2}$ differently with respect to $P_{tetO6inEGT2}$. Moreover, this is also important to check if the capacity of recruiting specific components involved in the activation of native *EGT2* promoter is conserved in the synthetic fusion proteins used with this experimental setup.

4.1.3 Results and Discussion

Doxycycline dependent activators were combined with the three target promoters (activators were transformed into α-strains while target promoters in *A*-strains, diploid strains combining different activator/promoter pairs were then obtained by yeast mating). The resulting strains were cultivated in the same conditions, exposed to the same doxycycline concentrations and reporter gene expression was measured by flow cytometry (see Fig.16-18).

The resulting dose-response curves indicate different ranges of expression between the analyzed constructs. Even though two different core promoters can have different inducibility due to their intrinsic specific properties [101, 102], the comparison between activation of $P_{tetO2inSIC1}$ and $P_{tetO6inEGT2}$ indicates that there is a non-linear relationship between the number of activators binding a promoter and the dynamical range of its response with roughly a 10-fold dynamical range for $P_{tetO2inSIC1}$ and up to 100-fold for $P_{tetO6inEGT2}$.

The presence of a weak native activator to $P_{tetO5inEGT2}$ has the effect of reducing the dynamical range of its response of about 2/5-fold. However, this is mostly due to the basal expression level which was raised 10/20-fold with respect to $P_{tetO6inEGT2}$. This indicates a unexpectedly strong role played by the remainings of native organization for the activation of this promoter. Hence, understanding the way fusion proteins mimic the natural activation process of the *EGT2* promoter appears not to be trivial because first of all, qualitative differences are noticeable by comparing the two versions of the Swi5p-based fusion proteins, and secondly, the observed effect is not consistent among the different target promoters tested. Even if functional, the fact that fusion proteins behave differently between each other is potentially due to their structure. Altered flexibility or steric hindrance influencing components of the transcriptional machinery could be reasons that explain why the fusion proteins used here behave differently between each other in the same promoter context and differently among different promoters.

The quantitative analysis of the data based on non-linear regression fitting to Hill equation (Eq.47) turned out to be less informative then expected because partially biased by the measure of V_{max}. The value of the maximally attainable output is not clearly

4.1 Context-Dependent Effects on Gene Regulation — Results

determinable. Even if the fitted curves are in most cases reaching a plateau at maximal doxycycline induction, the extrapolated V_{max} values can not be considered reliably due to scarcity of data in the high induction range. It is especially unclear if different constructs will tend to the same asymptotic value at induction tendent to infinity. This is a consequence of the fact that *in vivo* systems preferentially tend to have ranges of action far from saturation. Differently from *in vitro* systems, where the level of input (e.g. enzyme substrate or ligand that binds a receptor) can be raised to almost unlimited concentrations (except if substrate-induced receptor's aggregation or unfolding occurs), doxycycline concentrations higher than 10 µg/ml (= 20 µM) induce toxic effects in cells. Further, since K_D is defined in the Hill equation as to be the x-coordinate value corresponding to y-coordinate of $V_{max}/2$, its accuracy is subject to the same problem as for V_{max} determination.

As discussed in the section 'The Hill Equation', despite the fact that the Hill coefficient n_H is devoid of physical meaning and should be considered as a purely empirical value, its extrapolation by non-linear regression curve fitting to the presented experimental dose-response data is reliable. The Hill coefficient is a measure of sigmoidicity of the fitted curve and is commonly associated to the degree of cooperativity of the system. The data indicate that while the $P_{tetO2inSIC1}$ shows absence or negative cooperativity with respect to the different activators, a significant difference is observed between $P_{tetO5inEGT2}$ and $P_{tetO6inEGT2}$. The presence of a unique native activator binding to $P_{tetO5inEGT2}$ has the effect of reducing the cooperativity level compared to $P_{tetO6inEGT2}$. The observed effect hints for a change of promoter configuration induced by specific components recruited by the native Swi5p/Ace2p binding independently of external induction. This change in promoter reactivity potentially relies on a modified accessibility of activators to other binding sites. A similar argument could explain the values of n_H measured in $P_{tetO5inEGT2}$ with rtTA, rtetR-NLS-Swi5pABCD, rtetR-Swi5p and rtetR-Ace2p. Size, disposition and flexibility of fusion proteins' domains could influence the mechanism of activator binding. More precisely, it is plausible that the fusion proteins rtetR-Swi5p and rtetR-Ace2p are less capable of interacting favorably together with the native activator binding to $P_{tetO5inEGT2}$ and the recruited specific components compared to rtetR-NLS-Swi5pABCD. Since rtetR-Swi5p and rtetR-Ace2p are simply consisting of the whole activator sequence connected to a tetR domain, it's likely that their positioning along DNA binding site can be significantly different from the natural situation. This would result in unfavorable interactions between natural and synthetic activators which would compromise the process of gene activation and could explain their low Hill coefficients (compare Fig.16-18). While rtTA is not supposed to undergo gene specific interactions with native Swi5p/Ace2p due to its extraneous origin, and in addition its reduced size (335 amino acids) makes steric interactions unlikely, rtetR-NLS-Swi5pABCD could be a candidate for favorable interaction with native Swi5p/Ace2p in $P_{tetO5inEGT2}$. Due to its higher structural similarity to the natural Swi5p compared to rtetR-Swi5p, rtetR-NLS-Swi5pABCD could fit better the configuration of the promoter bound by a native activator thereby allowing the occurrence of positive interactions detected by the highest measured Hill coefficient (n_H = 1.25) (see Fig.17). This is also supported by the $P_{tetO2inSIC1}$ curves where rtetR-NLS-Swi5pABCD appears to be the only activator able of activation through non-negative cooperative binding (see Fig.16). This case could be an indication of negative steric interactions between activators that would correlate n_H values inversely to activator size (Ace2p is slightly larger than Swi5p, 770 and 709 amino acids, respectively) and is consistently observed over all three synthetic promoter measurements (compare Fig.16-18).

The most straightforward interpretation of the results is that native activators binding to $EGT2$ promoter have the capacity to alter its state and make it more responsive/accessible to further elements involved in its activation. This finds support by the values of last induction points (also indicated by V_{max} values) and part of the induction curves of $P_{tetO5inEGT2}$ having higher expression values than $P_{tetO6inEGT2}$ (see Fig.17 and Fig.18). A change in promoter configuration mediated by Swi5p/Ace2p such as acetylases recruitment and consequent loosening of chromatin structure could switch the $EGT2$ promoter to a state of higher activity. In contrast, according to this hypothesis, $P_{tetO6inEGT2}$ would be less accessible because of its higher chromatin condensation level and thereby each activator binding would contribute to its unloosening facilitating the subsequent activator binding. This represents a possible reason supporting the large Hill coefficient values measured for all activators binding to $P_{tetO6inEGT2}$. Moreover, the fact that in this case rtTA has highest n_H could be due to its small size (335 amino acids) and consequent facilitated access to the binding sites within chromatin structured promoter (see Fig.18).

4.1.4 Conclusions

The experiments carried out in this part allowed to gain insights in the study of gene activation based on synthetic components introduced in living yeast cells. Information could be obtained about cooperative binding of transcriptional activators to DNA and how promoters integrate the signal mediated by different activators. It resulted that context specific effects influencing regulation of the different studied systems are relevant for converting induction stimuli into genetic responses. This could in effect largely shape gene activation profiles based on the involved activation domain, and consequently the differences in the recruited molecular components, as well as the promoter sequence and the number of activator binding sites contained in it. Binding cooperativity could thereby influence gene activation profiles and be supported differently as a function of the different molecular components involved in the process.

The advantage of the genetic approach used in this work consists in taking into account the *in vivo* context specificity typically neglected in *in vitro* experiments. However, as a consequence of the important role of context specific effects, the use of enzyme kinetics tools to interpret gene activation curves resulted to be not fully appropriate. Even though the Hill equation allows to measure the sigmoidicity of the dose-response curve as a measure of binding cooperativity, the information about response's dynamical range and relative expression levels between different promoters as well as activator's apparent binding affinity K_D could not be retrieved with high reliability. Actually, both Hill and Adair equations commonly used in enzyme kinetics suppose relatively precise knowledge of V_{max}, which is, as explained earlier, more easily determinable when dealing with *in vitro* systems than with *in vivo* genetic systems. This relies on the fact that it is not known to what extent single transcriptional activators bound to a promoter contribute to its activation. As a comparison, if studying enzyme kinetics *in vitro* we can make sure that if one isolated subunit of an enzyme has a certain catalytic activity at saturating substrate concentrations, then the n-meric form of the enzyme under the same conditions will catalyze n-times the amount of substrate with respect to the single subunit. This is because such *in vitro* reactions are characterized by relatively

4.1 Context-Dependent Effects on Gene Regulation Results

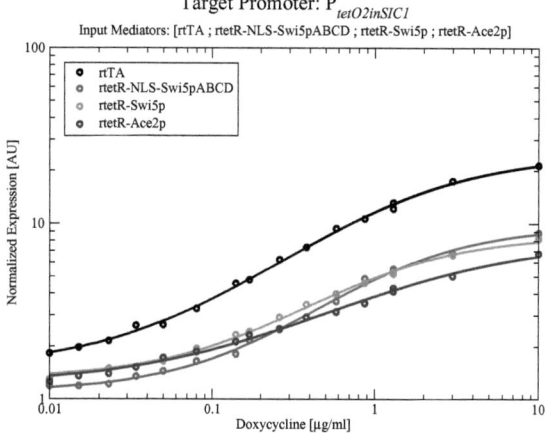

$P_{tetO2inSIC1}$	B	V_{max}	K_D	n_H
rtTA	1.50	23.43	1.38	0.86
rtetR-NLS-Swi5pABCD	1.10	8.70	1.35	1.00
rtetR-Swi5p	1.26	7.36	1.04	0.96
rtetR-Ace2p	1.21	6.88	1.90	0.73

Figure 16: $P_{tetO2inSIC1}$ driving expression of GFP under control of doxycycline dependent activators (rtTA, rtetR-NLS-Swi5pABCD, rtetR-Swi5p, rtetR-Ace2p). Saturation dose-response curves as a function of the applied doxycycline concentration are shown in a graphical representation in the upper plot while the lower table shows the results of non-linear regression fitting of the curves corresponding to the different constructs. The datasets were fitted with non-linear regression to Hill equation (Eq.47) to which a basal expression term (B, corresponding to the expression value measured at zero doxycycline) has been added and kept fixed during the fitting procedure. The resulting optimal parameters are reported in the table.

4.1 Context-Dependent Effects on Gene Regulation   Results

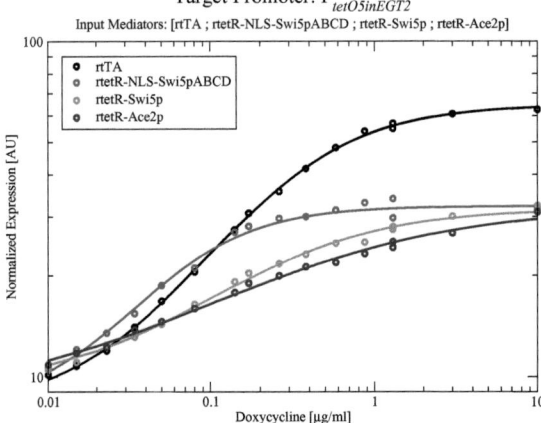

$P_{tetO5inEGT2}$	B	V_{max}	K_D	n_H
rtTA	8.02	56.76	0.26	1.05
rtetR-NLS-Swi5pABCD	8.04	24.18	0.06	1.25
rtetR-Swi5p	9.32	22.35	0.21	0.87
rtetR-Ace2p	8.54	23.52	0.31	0.60

Figure 17: $P_{tetO5inEGT2}$ driving expression of GFP under control of doxycycline dependent activators (rtTA, rtetR-NLS-Swi5pABCD, rtetR-Swi5p, rtetR-Ace2p). Saturation dose-response curves as a function of the applied doxycycline concentration are shown in a graphical representation in the upper plot while the lower table shows the results of non-linear regression fitting of the curves corresponding to the different constructs. The datasets were fitted with non-linear regression to Hill equation (Eq.47) to which a basal expression term (B, corresponding to the expression value measured at zero doxycycline) has been added and kept fixed during the fitting procedure. The resulting optimal parameters are reported in the table.

4.1 Context-Dependent Effects on Gene Regulation   Results

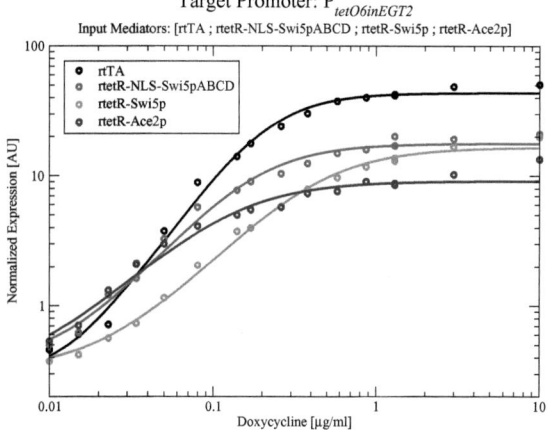

$P_{tetO6inEGT2}$	B	V_{max}	K_D	n_H
rtTA	0.26	43.35	0.20	1.88
rtetR-NLS-Swi5pABCD	0.35	17.40	0.16	1.59
rtetR-Swi5p	0.32	16.21	0.40	1.43
rtetR-Ace2p	0.28	8.89	0.12	1.34

Figure 18: $P_{tetO6inEGT2}$ driving expression of GFP under control of doxycycline dependent activators (rtTA, rtetR-NLS-Swi5pABCD, rtetR-Swi5p, rtetR-Ace2p). Saturation dose-response curves as a function of the applied doxycycline concentration are shown in a graphical representation in the upper plot while the lower table shows the results of non-linear regression fitting of the curves corresponding to the different constructs. The datasets were fitted with non-linear regression to Hill equation (Eq.47) to which a basal expression term (B, corresponding to the expression value measured at zero doxycycline) has been added and kept fixed during the fitting procedure. The resulting optimal parameters are reported in the table.

4.1 Context-Dependent Effects on Gene Regulation

well known thermodynamic constraints. As explained in the introduction concerning the hemoglobin example, if one subunit of hemoglobin transports one molecule of oxygen, the tetrameric hemoglobin molecule will transport four oxygen molecules. In contrast, if one transcriptional activator binding site, under very strong induction, activates a promoter such that x RNA molecules are transcribed, then by increasing to n number of activator binding sites in the promoter will not imply that $(n \cdot x)$ RNA molecules will be transcribed. It is actually more realistic to think that upon binding, an activator contributes to the recruitment and assembly of the RNA polymerase II machinery in order to initiate transcription and that the presence of multiple binding sites in a promoter simply increases this probability to some extent but most likely not linearly as it is the case for enzymes where multiple subunits contribute separately.

Hence, the results provided by these studies pointed out certain aspects of gene regulations that required further investigations. We realized that a new experimental approach together with appropriate tools to interpret our data were needed to understand in more depth the specific role of multiple binding sites present in gene regulatory sequences. So, we decided to turn to a more systematic approach mostly focused on varying the number of activator binding sites in promoters in a way that experimental outcomes could be analyzed using alternative methods less specific to enzyme kinetics. The approach adopted in the following sections aims to improve our knowledge regarding gene regulation at the level of the microscopic processes occurring in activator-DNA interaction in relation to binding cooperativity and at the same time to investigate the role of promoter specific effects in modulating signal transduction.

4.2 Systematic Study on the Effect of Multiple Activator Binding Sites in Promoters

4.2.1 Introduction

In order to access the mechanism of gene regulation concerning cooperative activator binding to DNA at the molecular interaction level, we developed an additional series of modulable-control genetic circuits in *Saccharomyces cerevisiae*. The principle of this new approach is to systematically increase the number of binding sites in the target promoter under study. In this way, under same induction conditions and using the appropriate mathematical tools for data interpretation, the outcome would provide information about the net effect gained by introduction of multiple activator binding sites. Moreover, parameters unaccessible *in vivo* would have smaller influence on the quantitative interpretation of the data.

Synthetically engineered gene modules that interact with a natural selected subsystem allow to experimentally observe how perturbations affect gene regulation in its natural context, thereby obtaining helpful information about relevant characteristics of the studied system [103, 104, 105, 106, 107, 1]. We therefore engineered a new set of gene modules in order to correlate activation input of gene regulatory sequences to reporter gene output signal. The the difference between these modules consisted principally of a variable number of binding sites for a common regulator. Consequently, properties of different gene modules could be expressed as a function of increasing regulator binding site number thereby representing an additional variable to describe the studied systems.

In the following sections we define three important categories of experiments based on the employed components used to characterize the system under study. This categorization (*full-synthetic*, *semi-synthetic* and *natural* systems) is not completely justified *per se* because the boundaries that separate the types of systems cannot be delimited unambiguously. We anyway use this notation accordingly to the following sections for the purpose of a better sectioning of the experimental design part and for facilitating understanding.

4.2 Systematic Study on the Effect of Multiple Activator Binding Sites in Promoters
Results

4.2.2 Full-Synthetic Systems

> **Motivation:** Both activator and promoter sequence play a role in converting induction stimuli into genetic response. Externally tunable activators can be used to gradually activate selected promoters and obtain saturation curves over a broad range of induction. Activation profiles of different promoters in relation to multiple activator binding sites would provide information about non-linearity and mechanisms underlying cooperativity in gene regulation.
> **Aims:** Understand the role of promoter sequences with respect to multiple activator binding sites and activator binding cooperativity. The influence of promoter sequences can be studied individually by replacing yeast's native activator binding sites with synthetic activator-operator elements minimally interacting with components of the cellular machinery. Synthetic tet-operators have been thereby inserted with increasing number in different target promoters aiming to study how different core promoter sequences support variable number of activator binding sites. Different mathematical tools and assumptions provide means to quantitatively interpret such experimental data.
> **Results:** Same input mediated by the same synthetic activator could be decoded differently and transduced into diverse gene activation responses based on the promoter context in which the binding occurs. Certain promoters support binding cooperativity while for others multiple binding sites enhance gene expression mainly as a consequence of a statistically favorable activator binding probability. A common feature among the studied promoters is represented by an important role played by low order binding events and their effect on converting promoters into a state of higher reactivity.

We define 'full-synthetic systems' those systems in which native yeast promoters were modified and the binding sites for endogenous activators were replaced by *tet*-operators. rtTA was used as activator to regulate the promoter under study. In this way the activity of the promoter of interest could be tuned externally through addition of doxycycline (proportional to the active rtTA) and the output could be red out by reporter gene expression (see Fig.19).

We arbitrarily define as full-synthetic the setup of these doxycycline titration experiments for the reason that the molecular partners involved in the binding reaction (rtTA, *tet*-operator and doxycycline) are extraneous to the metabolism of yeast. The rtTA–*tet*-operators system is in fact composed of bacterial and viral elements that minimally interact with cellular components of yeast cells. This system can thereby be efficiently used to regulate selected gene modules and individually focus on the properties of the core promoter sequence in integrating rtTA mediated inputs, however with the consequence of being poorly representative of the native activation process because of the missing components potentially recruited by native activators. More precisely, the use of rtTA could perhaps fail in recruiting particular components because its activation domain (VP16) is extraneous of yeast and thereby cannot integrate yeast's specific signals. On the other hand, as it will be detailed later, the degree of isolation of full-synthetic systems with respect to the rest of the cellular machinery has revealed to be of great advantage for testing the experimental setup proposed in this work as well as for the interpretation of the resulting data significantly devoid of unspecific effects that typically characterize *in vivo* experiments. Moreover, the binding of TetR to *tet*-operators has been described in crystallographic studies and has the potential to support binding cooperativity [108, 52].

We applied this method to study the core promoter region of three different genes: *EGT2* promoter (P_{EGT2}), *CYC1* promoter (P_{CYC1}) and *GAL1* promoter (P_{GAL1}). The

4.2 Systematic Study on the Effect of Multiple Activator Binding Sites in Promoters
Results

binding sites of the respective transcription factors were replaced by blocks of repetitive *tet*-operators present in variable number in the modified promoter and subsequently exposed to doxycycline induction gradients. Both the effect of different number of binding sites in the target promoter as well as distinct binding configurations of rtTA were systematically investigated.

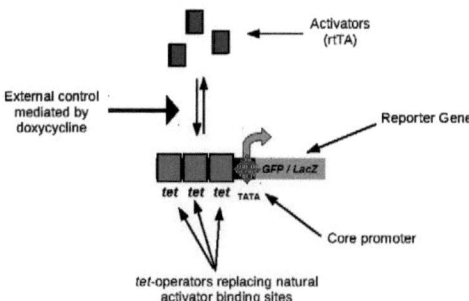

Figure 19: Principle of full-synthetic system to study the effect of the number of rtTA binding sites (*tet*-operators) on gene activation. The number of *tet*-operators was varied within the same promoter structure (the core promoter sequence was maintained unchanged) and cells containing the different promoters were exposed to the same induction conditions. In this way the effect of multiple activator binding could be studied and information about the role of core promoter sequences in interpreting activator mediated input could be retrieved.

4.2.2.1 System Description

$CYC1$ gene encodes the isoform 1 of the electron carrier cytochrome c. Its transcription is induced by oxygen, heme and lactate, and repressed by glucose. Oxygen response is mediated by the Hap1p transcription factor binding to one of two upstream activation sequences in P_{CYC1}. The second binding site is responsible for glucose repression which is mediated by Mig1p. P_{CYC1}-based constructs used in this section consist in the core promoter region of the $CYC1$ gene with different number of *tet*-operators (*tetO2* type) upstream of the CYC1-TATA region as described in [77].

P_{EGT2}, as explained in the previous section ('Context-Dependent Effects on Gene Regulation'), drives the expression of a cell cycle regulated endoglucanase. However, differently from the P_{EGT2}-based constructs used in the section 'Context-Dependent Effects on Gene Regulation' ($P_{tetO5inEGT2}$ and $P_{tetO6inEGT2}$) where the binding sites for the native regulators were substituted with *tet*-operators according to their natural dis-

4.2 Systematic Study on the Effect of Multiple Activator Binding Sites in Promoters
Results

position, here we simply maintained the *EGT* core promoter structure (192 bp upstream of the *EGT2* start codon) and inserted blocks of repetitive *tet*-operators (*tetO1* type) obtained synthetically.

The *GAL1* promoter (P_{GAL1}) controls the expression of the galactokinase involved in the metabolism of galactose. Together with other genes of the galactose metabolism network (*GAL1, GAL10, GAL7, GAL2*), it integrates signals mainly mediated by Gal4p, Gal80p and Gal3p to regulate genetic response to carbohydrates. P_{GAL1} contains four binding sites at close distance for the transcriptional activator Gal4p which were replaced by *tet*-operators (*tetO2* type) (see Fig.20). The operators that were not replaced by *tet*-operators in construct with less than 4 replacements were mutated in their consensus sequence thus abolishing Gal4p binding ($CGG[N]_{11}CCG \rightarrow GCC[N]_{11}GGC$).

The engineered systems containing *tet*-operators were activated by rtTA, whose expression was driven by the constitutive cycline 3 promoter (P_{CLN3}) and doxycycline used to control its affinity to their binding sites.

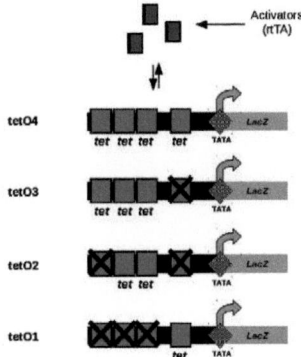

Figure 20: Replacements of native Gal4p binding sites by *tet*-operators in P_{GAL1}. Construct with less than 4 replacements were mutated in their consensus sequence ($CGG[N]_{11}CCG \rightarrow GCC[N]_{11}GGC$) which abolishes Gal4p binding (indicated by black crosses).

4.2.2.2 Data Interpretation Using the Hill Function

The full-synthetic method applied to the three different promoter systems ($P_{tetOinCYC1}$, $P_{tetOinEGT2}$, $P_{tetOinGAL1}$) clearly indicated a determining role of core promoter sequences in decoding signals mediated by a common activator. Comparing Fig.21,23,24 one immediately notices qualitative differences, especially in the way core promoters integrate signals mediated by a different number of activator binding sites.

Surprisingly, experimentally measured cooperativity showed quite substantial differences between the three analyzed systems indicating that the capacity of promoter sequences to modulate activator binding cooperativity doesn't depend on the properties

4.2 Systematic Study on the Effect of Multiple Activator Binding Sites in Promoters
Results

of the activator only but also on the target DNA sequence where it binds. Even if the activator used, rtTA, is not supposed to induce gene specific effects, the simple 'genetic environment' represented by the core promoter sequence was sufficient to determine different modes of gene activation. This reveals that promoter sequences play a very important role in interpreting activator mediated signals and are largely responsible for the emergence of complex mechanisms such as cooperativity in activator binding.

In $P_{tetOinCYC1}$ system (Fig.21), the degree of cooperativity measured by the Hill numbers strongly increases correlated to operator number. The Hill coefficient indicates negative cooperativity in the single operator construct, positive cooperativity with two and strong cooperativity with seven operators. The fact that systems with single activator binding sites display measured Hill coefficients different than one is common in situations were there are spatial heterogeneities which is the case for cellular environments. The present case of apparent negative cooperativity for single binding site constructs can rely on restricted promoter accessibility (e.g. a promoter switching between active and inactive states).

In contrast to $P_{tetOinCYC1}$, low cooperativity has been measured in $P_{tetOinEGT2}$ system (Fig.23). By methodically incrementing the number of rtTA binding sites from one to five the resulting cooperativity increases from negative to weakly positive. In addition to vary the number of activator binding sites in the $EGT2$ core promoter, the spatial binding configuration of rtTA to its binding sites has also been investigated. It has been shown and discussed how the three dimensional orientation of DNA binding proteins can have an influence on their binding mode [7, 109]. DNA binding site sequences spaced by 6 bp result in activators binding to opposite sites of the DNA helix. The protein-protein interaction that could potentially underly the effect of binding cooperativity would thereby be lost. If instead the binding site sequences are spaced by 10 bp then activators would bind the same side of the DNA helix on neighboring DNA helical turns thereby being in the same helical phasing (see Fig.22). This three dimensional binding configuration allows potential interaction between activators (e.g. protein-protein cross stabilization). The $P_{tetOinEGT2}$ series of promoters was made with tet-operators out of helical phasing. In order to verify that the low cooperativity measured with $P_{tetOinEGT2}$ system was not caused by the disruption of stabilizing or favorable contacts between adjacently bound rtTA molecules, the spacer distance has been increased in the two-operator construct from 6 bp to 10 bp (dashed black line, Fig.23) allowing activators to bind on same helical turns and to be compared to the corresponding construct out of helical phasing (red line, Fig.23). Visibly, helical phasing didn't influence positively activation of $P_{tetO2inEGT2}$, excluding protein-protein direct interaction as a cause for binding cooperativity.

Further, a crossing point between saturation curves with multiple and single activator binding sites is present in all receptor constructs, however, in the first case ($P_{tetOinCYC1}$, Fig.21) a very pronounced crossing is noticeable. The set of curves shown in Fig.21 hints for a graded curve distortion as a function of the number of activator binding sites: At low induction, the curves have a consistently higher activation level inversely proportional to the number of activator binding sites they contain while at higher induction the curve's trend inverts and gene expression correlates with number of binding sites. Moreover, the system appears to be limited: The number of binding sites introduced in the $CYC1$ core promoter can enhance gene activation and increase sensitivity up to a certain level only, indicating that increasing the number of binding sites cannot po-

4.2 Systematic Study on the Effect of Multiple Activator Binding Sites in Promoters
Results

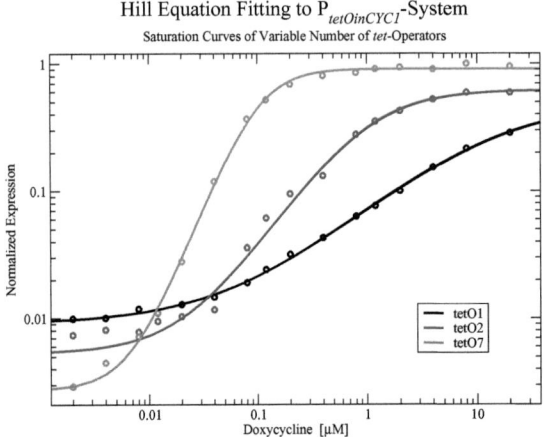

$P_{tetOinCYC1}$

	B	V_{max}	K_D	n_H
tetO1	$9.08 \cdot 10^{-3}$	0.49	13.11	0.75
tetO2	$5.32 \cdot 10^{-3}$	0.61	1.01	1.24
tetO7	$2.71 \cdot 10^{-3}$	0.91	0.10	2.12

Figure 21: Different number of *tet*-operators in P_{CYC1} driving expression of GFP under control of doxycycline dependent activator rtTA. The upper plot shows the graphical representation of the activation of target promoters containing variable number of *tet*-operators. The promoters are activated by rtTA and the saturation curves are represented as a function of the applied doxycycline concentration (proportional to the active rtTA). Expression values have been normalized to their maximal value. The resulting saturation curves were fitted with non-linear regression to the Hill equation and the extrapolated optimal parameters corresponding to the different constructs shown in the lower table (*tetO1*, *tetO2*, *tetO7* means that there are 1,2 and 7 operators in the $P_{tetOinCYC1}$ construct, respectively). The datasets were fitted with non-linear regression to Hill equation (Eq.47) to which a basal expression term (B, corresponding to the expression value measured at zero doxycycline) has been added and kept fixed during the fitting procedure.

4.2 Systematic Study on the Effect of Multiple Activator Binding Sites in Promoters
Results

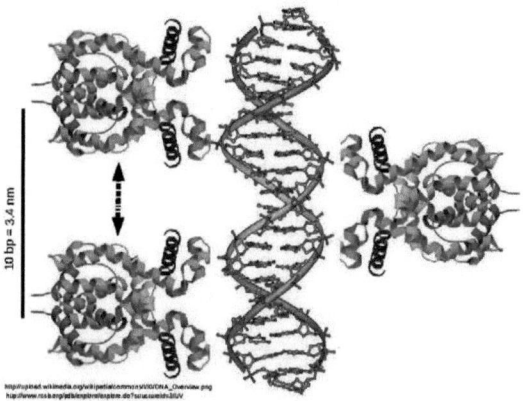

Figure 22: Schematic representation of DNA structural implications for regulators binding in different helical phasing. 10 bp spacing between binding sites causes activator to bind in helical phase (neighboring helical turns, same side of DNA helix) facilitating potential protein-protein interactions (black arrow). 6 bp spacing between binding sites causes activator to bind out of helical phase (opposite helical turns, opposite side of DNA helix) hindering potential protein-protein interactions.

tentiate activation over a certain threshold. The promoter containing seven activator binding sites appears to delimitate the maximal level of activation of the $P_{tetO in CYC1}$ system on one hand and system's sensitivity on the other hand. The maximal expression reached by this construct could in effect be determined reliably due to sufficient data in the saturating region. However, it is not clear to what maximal expression level the construct with one and two binding sites tend to. Due to the low amount of data points in their saturating region, it's hard to decide if assuming a common asymptotic limit for the three different constructs is justified. For the weaker constructs ($P_{tetO1inCYC1}$ and $P_{tetO2inCYC1}$) it is also difficult to interpret V_{max} values from non-linear regression analysis, according to which $P_{tetO1inCYC1}$ and $P_{tetO2inCYC1}$ asymptotically reach roughly 50% and 70% of $P_{tetO7inCYC1}$'s maximal activation, respectively. The boundaries of the $P_{tetOinCYC1}$ system are also suggested by the apparent dissociation constant K_D which decreases roughly 10-fold (reflecting saturation curves shifting to lower induction range) by doubling the number of binding sites from one to two, and decreases further again by 10-fold but this time with a binding site number increasing from two to seven. This means that the sensitivity of the system is limited in a similar fashion like its expression.

Activation of $P_{tetOinCYC1}$ system shows that differently from enzymatic systems where, as mentioned previously, a linear relation connects number of bound substrate molecules to converted product and hence to enzymatic activity, genes are not necessarily expressed proportionally to the number of activators bound to their promoters. This non-linear relationship connecting activator binding site number and gene expression level is also observable in $P_{tetOinEGT2}$ system (Fig.23). Another similarity between $P_{tetOinCYC1}$ and $P_{tetOinEGT2}$ system is that with more than three activator binding sites

4.2 Systematic Study on the Effect of Multiple Activator Binding Sites in Promoters
Results

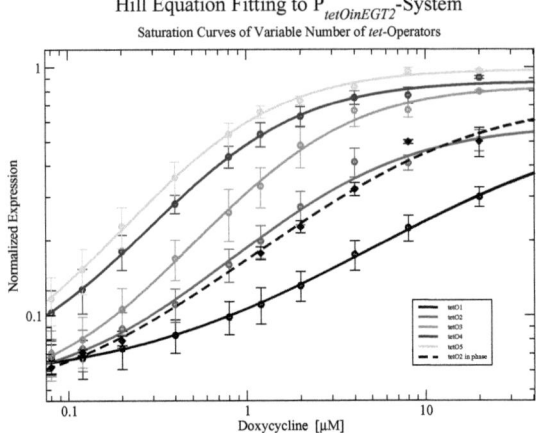

$P_{tetOinEGT2}$

	B	V_{max}	K_D	n_H
$tetO1$	$5.50 \cdot 10^{-2}$	0.57	29.62	0.69
$tetO2$	$4.98 \cdot 10^{-2}$	0.54	2.99	0.97
$tetO2$ in phase	$4.24 \cdot 10^{-2}$	0.71	6.84	0.80
$tetO3$	$4.81 \cdot 10^{-2}$	0.79	1.84	1.15
$tetO4$	$5.85 \cdot 10^{-2}$	0.82	0.93	1.16
$tetO5$	$5.29 \cdot 10^{-2}$	0.93	0.74	1.16

Figure 23: Different number of *tet*-operators in P_{EGT2} driving expression of GFP under control of doxycycline dependent activator rtTA. The upper plot shows the graphical representation of the activation of target promoters containing variable number of *tet*-operators. The promoters are activated by rtTA and the saturation curves are represented as a function of the applied doxycycline concentration (proportional to the active rtTA). The dashed line accounts for a distinct activator binding configuration where the activators bind on the same helical turn of the DNA (*tetO2 in phase*), thereby being at closer proximity with respect to the configuration in which they bind on opposite sides of the DNA helix (see Fig.22). Error bars are calculated with standard error of the mean from three experiments and expression values are normalized to their maximal value. The resulting saturation curves were fitted with non-linear regression to the Hill equation (Eq.47) to which a basal expression term (B, corresponding to the expression value measured at zero doxycycline) has been added and kept fixed during the fitting procedure. The extrapolated optimal parameters corresponding to the different constructs are shown in the lower table (*tetO1-tetO5* means that there are 1-5 operators in the $P_{tetOinEGT2}$ construct).

4.2 Systematic Study on the Effect of Multiple Activator Binding Sites in Promoters
Results

the system response changes very slightly compared to differences implied by a variable binding site number between one and three. While the Hill coefficients do not increase significantly further after a limit represented by the construct containing three operators, differences in K_D values between constructs containing n binding sites and constructs with $n+1$ binding sites decrease with increasing n, indicating a limit in gene response's sensitivity. In other words, by increasing number of operators from n to $n+1$, saturation curves shift to lower induction range. The larger n the weaker the shift.

The crossing point between single and multiple operators is a common feature shared between $P_{tetOinCYC1}$ and $P_{tetOinEGT2}$ and is also present in $P_{tetOinGAL1}$ (see Fig.24). This characteristic was consistent among all three constructs and is due to the large expression values at low induction ranges typical of single tet-operator constructs. This feature is especially relevant in the context of small regulatory networks and the role of basal expression for stochastic transitions in bistable systems (see 'Part III'). In contrast, $P_{tetOinGAL1}$ showed a different behavior with respect to the other constructs regarding the apparently absent upper bound delimiting maximal expression reachable by the system (at least up to four binding sites). From two to four binding sites the equal spacing between saturation curves suggests that this system behaves more similarly to enzymatic systems than the two other ones. $P_{tetOinGAL1}$ presents similarity to enzymatic systems also because its maximal expression was reached at maximal induction. With exception to the single operator construct, expression values are scaled proportionally to the number of operators. However, the striking particularity of this system relies on the qualitative difference observable between single *and* multiple binding site constructs. Interestingly, the data showed that the addition of a second activator binding site to the system principally influences the responsiveness of the promoter by making it more reactive at low induction ranges (mainly a large K_D shift to lower values). This reveals unexpected mechanism that can convert a linearly increasing variable such as the number of activator binding sites into a complex response pattern capable of displaying non-linear changes in the mode of activation. Hence, the saturation curves of this system displayed hybrid features common to both gene and enzyme based systems.

The results evaluated with Hill fitting indicated an important role played by low order binding events as a common feature between all the analyzed promoter systems. We consequently applied a different mathematical tool to interpret the same datasets. This allowed us to obtain more precise information about the quantitative contribution provided by each different binding event.

4.2 Systematic Study on the Effect of Multiple Activator Binding Sites in Promoters
Results

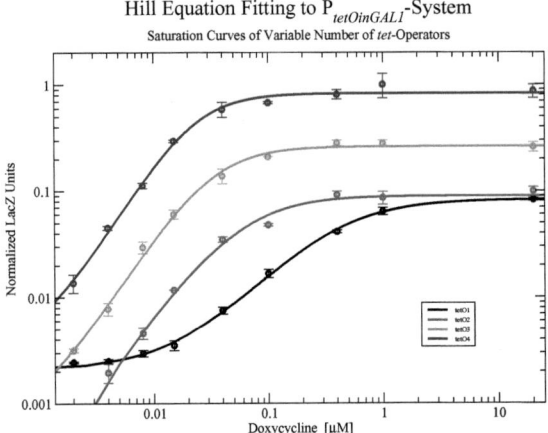

$P_{tetOinGAL1}$	B	V_{max}	K_D	n_H
tetO1	$2.07 \cdot 10^{-3}$	$8.05 \cdot 10^{-2}$	0.38	1.17
tetO2	$-3.69 \cdot 10^{-4}$	$8.98 \cdot 10^{-2}$	$6.12 \cdot 10^{-2}$	1.38
tetO3	$6.22 \cdot 10^{-4}$	0.26	$3.23 \cdot 10^{-2}$	1.65
tetO4	$3.50 \cdot 10^{-3}$	0.82	$2.22 \cdot 10^{-2}$	1.78

Figure 24: Different number of *tet*-operators in P_{GAL1} driving expression of LacZ under control of doxycycline dependent activator rtTA. Here the natural binding sites for Gal4p have been replaced by *tet*-operators (see Fig.20). The upper plot shows the graphical representation of the activation of target promoters containing variable number of *tet*-operators. The promoters are activated by rtTA and the saturation curves are represented as a function of the applied doxycycline concentration (proportional to the active rtTA). Error bars are calculated by standard deviation from two experiments and expression values are normalized to their maximal value. The resulting saturation curves were fitted with non-linear regression to the Hill equation (Eq.47) to which a basal expression term (B, corresponding to the expression value measured at zero doxycycline) has been added and kept fixed during the fitting procedure. The extrapolated optimal parameters corresponding to the different constructs are shown in the lower table (*tetO1-tetO4* means that 1-4 *tet*-operators are replacing native activator binding sites in the $P_{tetOinGAL1}$ construct, according to Fig.20).

4.2 Systematic Study on the Effect of Multiple Activator Binding Sites in Promoters
Results

4.2.2.3 Data Interpretation Using the Adair Equation

4.2.2.3.1 Sequential Non-linear Curve Fitting

The same datasets have been analyzed with help of Adair equations (enzyme-kinetics-Adair equation (Eq.31) and gene-regulation-Adair equation (Eq.38)). The fitting procedure has revealed not to be trivial because subject to overfitting. The major problem was however to be able to discern between the Adair equations the one that suites best the different datasets. Actually it is not clear if the systems described in this section are more efficiently represented being treated as enzymes (with different V_{max} values for constructs with variable number of binding sites) or to consider that single and multiple activators have common activation potential (with same V_{max} values for constructs with variable number of binding sites).

We first used a sequential approach to fit our experimental data obtained by exposing constructs of variable operator number to doxycycline gradients. Within a set of promoter constructs with increasing number of *tet*-operators, we iteratively extrapolated binding parameters that were kept fixed for next fitting procedure. Actually, Adair equation for $n + 1$ binding sites contains one more term that describes the last binding step in the series of n steps and also one additional unknown parameter with respect to the one for n binding sites. Since Adair equations assume conserved effects of bound activators on subsequent binding events, promoter constructs with increasing number of *tet*-operators can principally be used for sequential non-linear regression fitting with a single unknown parameter at a time. This would in fact reduce the risk of overfitting typical when dealing with equations consisting of many unknown parameters, but with the drawback of propagating errors through the datasets. The single operator construct was thereby used to extrapolate the parameter describing the first binding event. Since no binding cooperativity issues are considered in a binding process involving single ligands, systems with one activator binding sites provide information related to the binding affinity (K) of rtTA–*tet*-operator in the particular context represented by the promoter in which the binding occurs. The binding affinity K for the first activator binding is considered maintained and subsequent binding is influenced by a factor c_1, so fitting the Adair equation for two binding sites optimizes the best-fit value of c_1 assuming K constant. Similarly, the extrapolated c_1 value is held constant for fitting data of the construct containing three *tet*-operators and c_2 value is optimized for best non-linear curve fitting and so on. Due to problems in determining the expression range of different constructs, an additional variable parameter (S), acting as a scaling factor, multiplies the Adair equation in order to scale its output range. A second variable parameter (B) is further added to the equation to account for basal expression. Optimally fitting equations were further used to compute logarithmic sensitivity and the maximum point of the resulting function reported with the notation $max_x(\varepsilon)$ and $max_y(\varepsilon)$ indicating its coordinates corresponding to induction point and extent of maximal logarithmic sensitivity, respectively.

As it can be seen in Fig.27-30, the most relevant contributions in terms of promoter activation are encountered by adding operators to promoters containing few binding sites. The major contribution in $P_{tetOinGAL1}$ and $P_{tetOinEGT2}$ system results to be due to the second binding event, even if the differences for the latter case are less pronounced.

4.2 Systematic Study on the Effect of Multiple Activator Binding Sites in Promoters
Results

Further, the $P_{tetOinCYC1}$ system has also been analyzed (only partially, since the the iterative fitting procedure couldn't be applied between constructs of two and seven binding sites) (see Fig.25-26). The extrapolated values for this construct are roughly situated in between the two other ones. The apparent affinity measured by K indicates increasing activator binding strength in the following order: $P_{tetOinEGT2} < P_{tetOinCYC1} < P_{tetOinGAL1}$. $P_{tetOinGAL1}$ being about ten times stronger and $P_{tetOinCYC1}$ twice as strong as $P_{tetOinEGT2}$. Moreover, in $P_{tetOinCYC1}$ the enhancement factor for the second binding increases about 10/15-fold its apparent affinity with respect to the first, while for $P_{tetOinEGT2}$ about 2/10-fold and for $P_{tetOinGAL1}$ around 200-fold.

An important issue concerning the fitting procedure described in this section is revealed in Fig.26 where the same dataset has been fitted with two differently seeded non-linear regression procedures. The fitting algorithm consists in fact of iteratively improving the fitted function by varying parameters initially guessed up to a point of sufficient accuracy with respect to the dataset (Levenberg-Marquardt algorithm). This shows that even for relatively small equations (Adair equation for two binding sites) and few parameters, the choice of initial parameters can result in significantly evaluated parameters being trapped in local suboptimal points of the parameter space. In other words, the choice of two different initial guesses for the fitting procedures can give rise to divergent outcomes. This inconvenience can results in a drastic divergence between fitted datasets (compare dashed and continuous red lines in Fig.26) pointing out the high sensitivity of the approach. Furthermore, this could bias data interpretation especially in case of the iterative fitting procedure used here where parameters are extrapolated first and kept constant for the subsequent dataset analysis. Moreover, fitting data of multiple binding sites systems is especially problematic because the many terms composing their functions can compensate for parameters that were misestimated in previous fitting steps inducing error propagation through the whole dataset analysis.

The problem of suboptimal solutions remains anyway minimal when fitting datasets of low binding site number. In this case in fact a few trials of initial guesses can ensure that the system converges to same optimal solutions with most of the initially guessed parameters. In contrast, higher order terms present in the Adair equation formulated for many binding sites cause strong parameter cross compensation. Consequently, the higher order parameters evaluated in the fitting procedure result highly insignificant and can assume almost any value without having consequences for the outcome of the fitting. Any of these curves, even if subject to overfitting, can anyway be used to compute the logarithmic derivative and thereby being interpreted according to the point of maximal sensitivity. Even if such results are of lower informational content about the underlying physical process than Adair equations parameters, the method resulted to be consistent among different datasets and assumptions related to the function used for fitting as well as towards problems of local minima trapping.

An alternative in attempt to improve understanding of the analyzed datasets, though without solving the overfitting problems, is to avoid potential error propagation due to the cross influence between datasets consequent of the iterative procedure used in this section and carry out a fitting procedure by setting free all parameters defining each curve of the different systems.

4.2 Systematic Study on the Effect of Multiple Activator Binding Sites in Promoters

Results

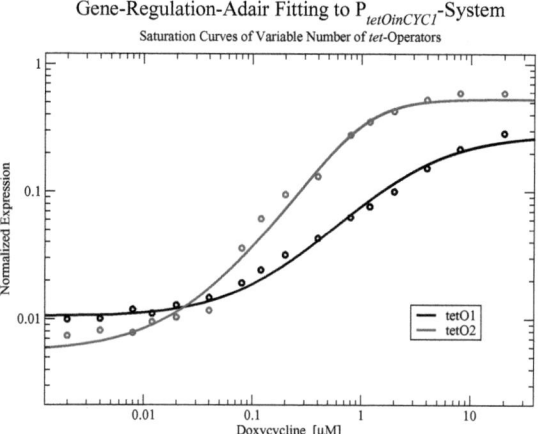

$P_{tetOinCYC1}$	S	B	K	c_1	$max_x(\varepsilon)$	$max_y(\varepsilon)$
tetO1	0.27	$1.03 \cdot 10^{-2}$	0.32		0.60	0.68
tetO2	0.49	$5.31 \cdot 10^{-3}$	0.32	9.72	0.24	0.99

Figure 25: $P_{tetOinCYC1}$ sequential fitting to gene-regulation-Adair equation (Eq.38). The fitting procedure is carried out by sequentially determining binding parameters that are kept fixed for successive fitting. K represents the apparent affinity of the rtTA to tet-operators (considered to be an association binding constant). c_1 is the enhancement factor that modulates second binding (positively in this case). $max_x(\varepsilon)$ and $max_y(\varepsilon)$ indicate the coordinates of the induction point of maximal sensitivity according to the corresponding fitted function. A basal expression term B added to the Adair equation, and a scaling factor S that multiplies it are additional free parameters used during the fitting procedure. The resulting optimal parameters are reported in the table.

4.2 Systematic Study on the Effect of Multiple Activator Binding Sites in Promoters
Results

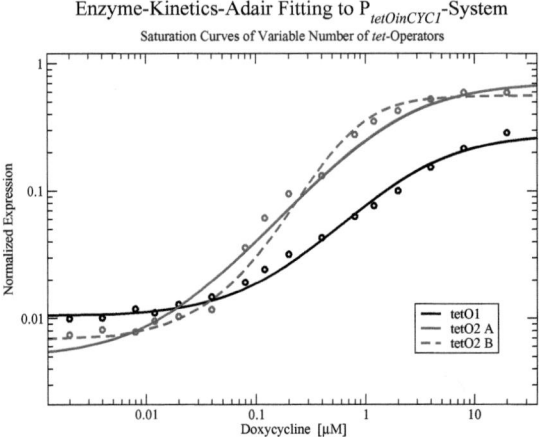

$P_{tetOinCYC1}$

	S	B	K	c_1	$max_x(\varepsilon)$	$max_y(\varepsilon)$
tetO1	0.27	$1.03 \cdot 10^{-2}$	0.32		0.60	0.68
tetO2 A	0.58	$4.99 \cdot 10^{-3}$	0.32	0.11	0.15	0.84
tetO2 B	0.29	$6.28 \cdot 10^{-3}$	0.32	15.80	0.25	1.18

Figure 26: $P_{tetOinCYC1}$ sequential fitting to enzyme-kinetics-Adair equation (Eq.31). The fitting procedure is carried out by sequentially determining binding parameters that are kept fixed for successive fitting. K represents the apparent affinity of the rtTA to *tet*-operators (considered to be an association binding constant). c_1 is the enhancement factor that modulates second binding. $max_x(\varepsilon)$ and $max_y(\varepsilon)$ indicate the coordinates of the induction point of maximal sensitivity according to the corresponding fitted function. A basal expression term B added to the Adair equation, and a scaling factor S that multiplies it are additional free parameters used during the fitting procedure. The resulting optimal parameters are reported in the table. Continuous and dashed red curves (*tetO2 A* and *tetO2 B*, respectively) represent the same dataset for rtTA binding to two *tet*-operators in which the initial guesses seeding the fitting procedure are different. This emphasizes the risk of potential local optimum trapping during fitting.

4.2 Systematic Study on the Effect of Multiple Activator Binding Sites in Promoters
Results

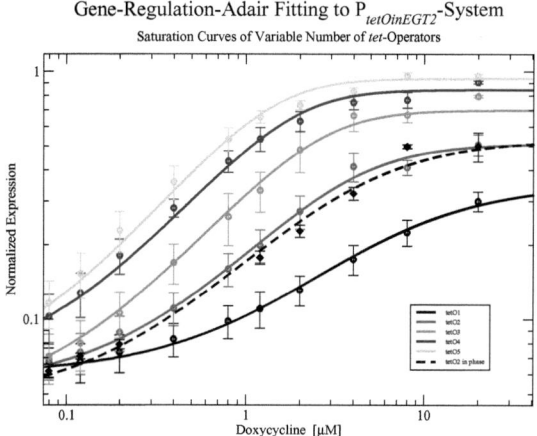

Gene-Regulation-Adair Fitting to $P_{tetOinEGT2}$-System
Saturation Curves of Variable Number of *tet*-Operators

$P_{tetOinEGT2}$

	S	B	K	c_1	c_2	c_3	c_4	$max_x(\varepsilon)$	$max_y(\varepsilon)$
tetO1	0.30	6.06·10^{-2}	0.17					2.41	0.42
tetO2	0.46	5.33·10^{-2}	0.17	2.06				1.05	0.56
tetO2 in phase	0.48	4.71·10^{-2}	0.17	0.41				0.92	0.56
tetO3	0.65	4.52·10^{-2}	0.17	2.06	1.29			0.61	0.70
tetO4	0.78	6.13·10^{-2}	0.17	2.06	1.29	2.36·10^{-5}		0.50	0.70
tetO5	0.88	5.58·10^{-2}	0.17	2.06	1.29	2.36·10^{-5}	1 (any value)	0.37	0.74

Figure 27: $P_{tetOinEGT2}$ sequential fitting to gene-regulation-Adair equation (Eq.38). The fitting procedure is carried out by sequentially determining binding parameters that are kept fixed for successive fitting. K represents the apparent affinity of the rtTA to *tet*-operators (considered to be an association binding constant). c_i is the enhancement factor that modulates i^{th} binding. *tetO2 in phase* represents a construct in which the activator binding sites are spaced in a way to allow closer contact to adjacent rtTA bound to the DNA. Higher order parameters determined for constructs with more than two binding sites are degenerated. $max_x(\varepsilon)$ and $max_y(\varepsilon)$ indicate the coordinates of the induction point of maximal sensitivity according to the corresponding fitted function. A basal expression term B added to the Adair equation, and a scaling factor S that multiplies it are additional free parameters used during the fitting procedure. The resulting optimal parameters are reported in the table.

4.2 Systematic Study on the Effect of Multiple Activator Binding Sites in Promoters
Results

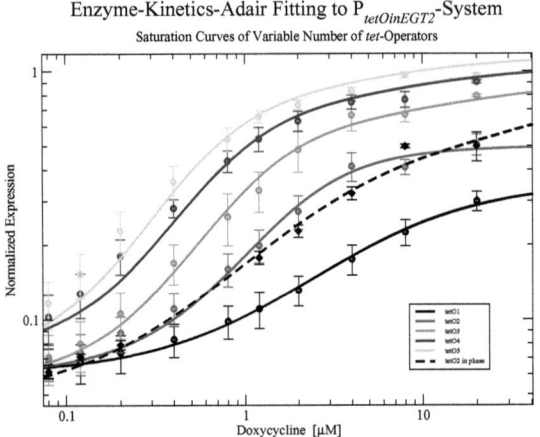

$P_{tetOinEGT2}$	S	B	K	c_1	c_2	c_3	c_4		$max_x(\varepsilon)$	$max_y(\varepsilon)$
$tetO1$	0.30	$6.06 \cdot 10^{-2}$	0.17						2.41	0.42
$tetO2$	0.22	$5.87 \cdot 10^{-2}$	0.17	9.59					1.02	0.72
$tetO2$ in phase	0.30	$5.04 \cdot 10^{-2}$	0.17	1.84					1.58	0.63
$tetO3$	0.29	$5.32 \cdot 10^{-2}$	0.17	9.59	$9.9 \cdot 10^{-2}$				0.56	0.86
$tetO4$	0.33	$6.99 \cdot 10^{-2}$	0.17	9.59	$9.9 \cdot 10^{-2}$	$5.81 \cdot 10^{-5}$			0.41	0.85
$tetO5$	0.36	$6.30 \cdot 10^{-2}$	0.17	9.59	$9.9 \cdot 10^{-2}$	$5.81 \cdot 10^{-5}$	1 (any value)		0.31	0.90

Figure 28: $P_{tetOinEGT2}$ sequential fitting to enzyme-kinetics-Adair equation (Eq.31). The fitting procedure is carried out by sequentially determining binding parameters that are kept fixed for successive fitting. K represents the apparent affinity of the rtTA to tet-operators (considered to be an association binding constant). c_i is the enhancement factor that modulates i^{th} binding. $tetO2$ in phase represents a construct in which the activator binding sites are spaced in a way to allow closer contact to adjacent rtTA bound to the DNA. Higher order parameters determined for constructs with more than two binding sites are degenerated. $max_x(\varepsilon)$ and $max_y(\varepsilon)$ indicate the coordinates of the induction point of maximal sensitivity according to the corresponding fitted function. A basal expression term B added to the Adair equation, and a scaling factor S that multiplies it are additional free parameters used during the fitting procedure. The resulting optimal parameters are reported in the table.

4.2 Systematic Study on the Effect of Multiple Activator Binding Sites in Promoters
Results

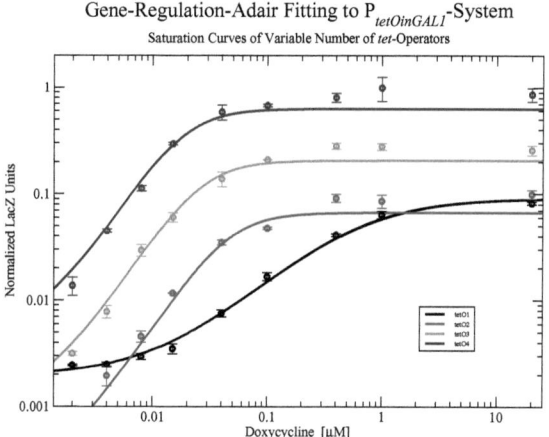

$P_{tetOinGAL1}$	S	B	K	c_1	c_2	c_3	$max_x(\varepsilon)$	$max_y(\varepsilon)$
tetO1	$8.96 \cdot 10^{-2}$	$1.93 \cdot 10^{-3}$	1.70				$8.54 \cdot 10^{-2}$	0.75
tetO2	$6.71 \cdot 10^{-2}$	$4.54 \cdot 10^{-9}$	1.70	188.51			$1.06 \cdot 10^{-2}$	1.49
tetO3	0.21	$5.81 \cdot 10^{-4}$	1.70	188.51	$4.08 \cdot 10^{-8}$		$6.83 \cdot 10^{-3}$	1.48
tetO4	0.63	$3.10 \cdot 10^{-3}$	1.70	188.51	$4.08 \cdot 10^{-8}$	1 (any value)	$5.15 \cdot 10^{-3}$	1.46

Figure 29: $P_{tetOinGAL1}$ sequential fitting to gene-regulation-Adair equation (Eq.38). The fitting procedure is carried out by sequentially determining binding parameters that are kept fixed for successive fitting. K represents the apparent affinity of the rtTA to tet-operators (considered to be an association binding constant). c_i is the enhancement factor that modulates i^{th} binding. Higher order parameters determined for constructs with more than two binding sites are degenerated. $max_x(\varepsilon)$ and $max_y(\varepsilon)$ indicate the coordinates of the induction point of maximal sensitivity according to the corresponding fitted function. A basal expression term B added to the Adair equation, and a scaling factor S that multiplies it are additional free parameters used during the fitting procedure. The resulting optimal parameters are reported in the table.

4.2 Systematic Study on the Effect of Multiple Activator Binding Sites in Promoters
Results

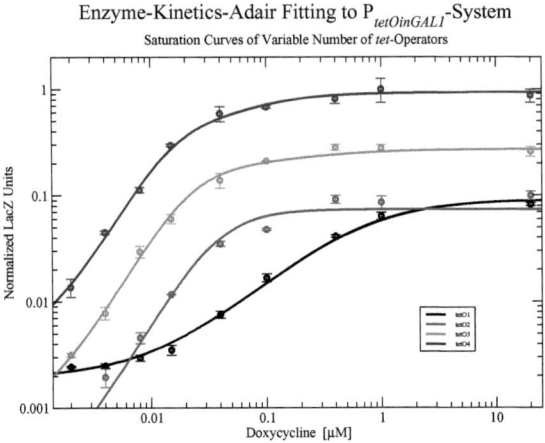

$P_{tetOinGAL1}$	S	B	K	c_1	c_2	c_3	$max_x(\varepsilon)$	$max_y(\varepsilon)$
tetO1	$8.96 \cdot 10^{-2}$	$1.93 \cdot 10^{-3}$	1.70				$8.54 \cdot 10^{-2}$	0.75
tetO2	$3.7 \cdot 10^{-2}$	$1.4 \cdot 10^{-6}$	1.70	247.53			$7.88 \cdot 10^{-3}$	1.66
tetO3	$9.02 \cdot 10^{-2}$	$6.21 \cdot 10^{-4}$	1.70	247.53	$3.65 \cdot 10^{-2}$		$5.98 \cdot 10^{-3}$	1.64
tetO4	0.23	$3.47 \cdot 10^{-3}$	1.70	247.53	$3.65 \cdot 10^{-2}$	3.15	$4.94 \cdot 10^{-3}$	1.61

Figure 30: $P_{tetOinGAL1}$ sequential fitting to enzyme-kinetics-Adair equation (Eq.31). The fitting procedure is carried out by sequentially determining binding parameters that are kept fixed for successive fitting. K represents the apparent affinity of the rtTA to *tet*-operators (considered to be an association binding constant). c_i is the enhancement factor that modulates i^{th} binding. Higher order parameters determined for constructs with more than two binding sites are degenerated. $max_x(\varepsilon)$ and $max_y(\varepsilon)$ indicate the coordinates of the induction point of maximal sensitivity according to the corresponding fitted function. A basal expression term B added to the Adair equation, and a scaling factor S that multiplies it are additional free parameters used during the fitting procedure. The resulting optimal parameters are reported in the table.

4.2 Systematic Study on the Effect of Multiple Activator Binding Sites in Promoters
Results

4.2.2.3.2 Global Non-linear Curve Fitting of Single Datasets

An alternative to the analysis carried out in the previous section is to globally fit single datasets without fixing any parameters from other curves and by constraining parameters of the fitted function to assume positive (physically relevant) values. To verify that Adair equation analysis presented in the previous section was not strongly suffering local minima trapping of low binding site constructs' fitting, with the implied error propagation consequences, we repropose the same analysis attempting to perform non-linear curve fitting by extrapolating all parameters at once for each system. This allows to check agreement with respect to the analysis performed in the previous section and assesses error propagation effects related to it but has the disadvantage of being even more prone to overfitting since a larger number of parameters is varied. Consequently, this analysis is only reliable with constructs containing maximally three binding sites, otherwise the fitting procedure degenerates and the variable parameters cross compensate becoming independent on the fitting procedure (the values that higher order parameters can assume range very widely without affecting the fitting).

With this approach we obtained for each curve a different measure of the same parameter, which could be compared to the unique value measured in the sequential fitting approach. It can be noticed that the set of apparent affinities K measured here within the same system displays values that can be more or less similar to the ones measured in the previous section depending on the Adair equation type used for fitting. In general, with respect to the sequential fitting approach, K values are consistently overestimated using enzyme-kinetics-Adair equation and underestimated using gene-regulation-Adair equation. Actually, if the K values measured by the two approaches are reasonably matching regarding the $P_{tetOinCYC1}$ system (compare Fig.25/26 with Fig.31/32), for the two other promoters a difference can be observed depending on the assumption made to fit the dataset (which type of Adair equation is used). Interestingly, using gene-regulation-Adair equation for fitting curves of the $P_{tetOinEGT2}$ constructs and enzyme-kinetics-Adair equation for fitting curves of the $P_{tetOinGAL1}$ constructs consistently corresponds to the K measures resulting from sequential fitting. (compare Fig.27/28 with Fig.33/34 and Fig.29/30 with Fig.35/36). This supports the earlier statements according to which $P_{tetOinGAL1}$ system shares more features typical of enzymatic systems than $P_{tetOinEGT2}$, the latter exhibiting a behavior more representative to the assumption made when deriving gene-regulation-Adair equation.

By comparing global and sequential fitting analysis, it results that both fitting procedures are in relative good agreement concerning apparent binding affinities and enhancement factors of first activator binding events. The sensitivity range of different constructs can be determined quite reliably in accord to the outcome presented in the previous section. The broadest dynamical range of activation determined by the apparent rtTA–tet-operator affinity situates $P_{tetOinCYC1}$ and $P_{tetOinGAL1}$ systems at activation ranges 2-fold and 10-fold more concentration sensitive with respect to the $P_{tetOinEGT2}$, respectively (compare Fig.25/26 with Fig.27/28 and Fig.29/30 from sequential fitting as well as Fig.31/32 with Fig.33 and Fig.36 from global fitting of single datasets). For all promoter systems the first binding event represents the most important contribution to enhance further activator binding. This effect could be quantified and revealed a different contribution of the first binding to enhance the second among the studied promoter systems. The major binding enhancement was measured for $P_{tetOinGAL1}$ where

4.2 Systematic Study on the Effect of Multiple Activator Binding Sites in Promoters
Results

200/250-fold higher apparent affinity resulted as a consequence of strong cooperativity between the two first binding activators (see Fig.30,36). The corresponding binding site affinity enhancement for $P_{tetOinEGT2}$ and $P_{tetOinCYC1}$ was about 2-fold and 10/20-fold, respectively (see Fig.27,33 and Fig.25/26,31/32, respectively).

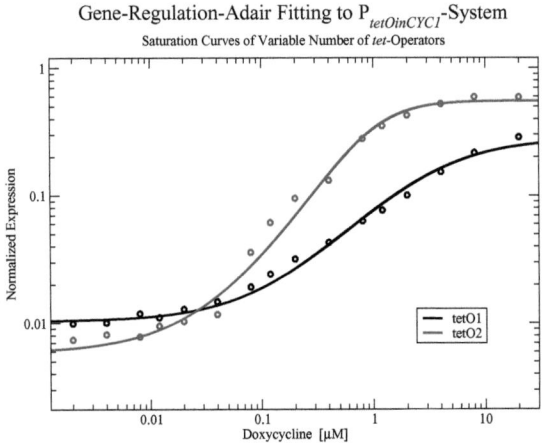

$P_{tetOinCYC1}$

	S	B	K	c_1	$max_x(\varepsilon)$	$max_y(\varepsilon)$
tetO1	0.27	$1.03 \cdot 10^{-2}$	0.32		0.60	0.68
tetO2	0.55	$5.87 \cdot 10^{-3}$	0.22	21.68	0.28	1.09

Figure 31: $P_{tetOinCYC1}$ single dataset fitting to gene-regulation-Adair equation (Eq.38). The fitting procedure is carried out by varying all binding parameters for each curve separately. The variable parameters are restricted to be greater than zero. K represents the apparent affinity of the rtTA to *tet*-operators (considered to be an association binding constant). c_1 is the enhancement factor that modulates second binding (positively in this case). $max_x(\varepsilon)$ and $max_y(\varepsilon)$ indicate the coordinates of the induction point of maximal sensitivity according to the corresponding fitted function. A basal expression term B added to the Adair equation, and a scaling factor S that multiplies it are additional free parameters used during the fitting procedure. The resulting optimal parameters are reported in the table.

4.2 Systematic Study on the Effect of Multiple Activator Binding Sites in Promoters
Results

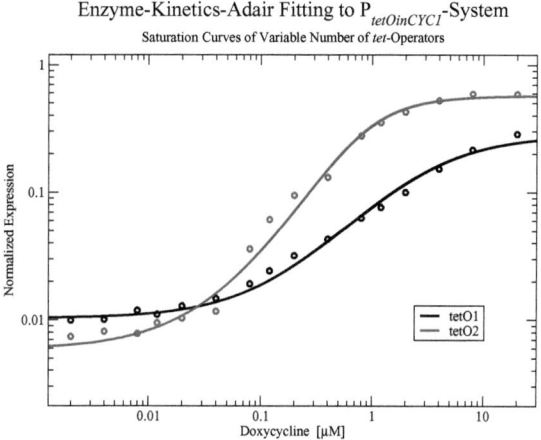

$P_{tetOinCYC1}$

	S	B	K	c_1	$max_x(\varepsilon)$	$max_y(\varepsilon)$
tetO1	0.27	1.03·10^{-2}	0.32		0.60	0.68
tetO2	0.28	5.92·10^{-3}	0.41	8.87	0.24	1.10

Figure 32: $P_{tetOinCYC1}$ single dataset fitting to enzyme-kinetics-Adair equation (Eq.31). The fitting procedure is carried out by varying all binding parameters for each curve separately. The variable parameters are restricted to be greater than zero. K represents the apparent affinity of the rtTA to *tet*-operators (considered to be an association binding constant). c_1 is the enhancement factor that modulates second binding (positively in this case). $max_x(\varepsilon)$ and $max_y(\varepsilon)$ indicate the coordinates of the induction point of maximal sensitivity according to the corresponding fitted function. A basal expression term B added to the Adair equation, and a scaling factor S that multiplies it are additional free parameters used during the fitting procedure. The resulting optimal parameters are reported in the table.

4.2 Systematic Study on the Effect of Multiple Activator Binding Sites in Promoters
Results

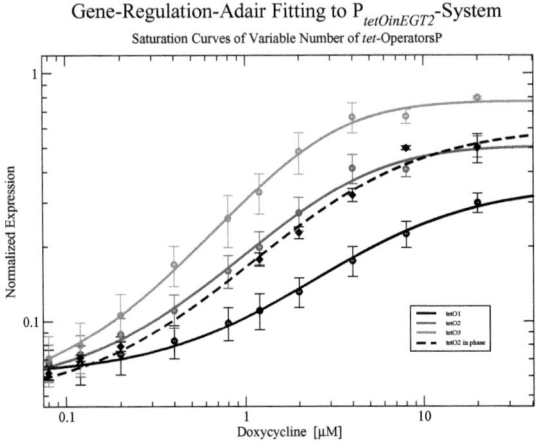

$P_{tetOinEGT2}$	S	B	K	c_1	c_2		$max_x(\varepsilon)$	$max_y(\varepsilon)$
tetO1	0.30	6.06·10^{-2}	0.17				2.41	0.42
tetO2	0.46	5.23·10^{-2}	0.18	1.41			0.98	0.57
tetO3	0.73	4.77·10^{-2}	0.13	2.80	7.07·10^{-4}		0.72	0.71
tetO2 in phase	0.53	4.80·10^{-2}	0.14	0.32			1.07	0.57

Figure 33: $P_{tetOinEGT2}$ single dataset fitting to gene-regulation-Adair equation (Eq.38). The fitting procedure is carried out by varying all binding parameters for each curve separately. The variable parameters are restricted to be greater than zero. K represents the apparent affinity of the rtTA to tet-operators (considered to be an association binding constant). c_i is the enhancement factor that modulates i^{th} binding. tetO2 in phase represents a construct in which the activator binding sites are spaced in a way to allow closer contact to adjacent rtTA bound to the DNA. $max_x(\varepsilon)$ and $max_y(\varepsilon)$ indicate the coordinates of the induction point of maximal sensitivity according to the corresponding fitted function. A basal expression term B added to the Adair equation, and a scaling factor S that multiplies it are additional free parameters used during the fitting procedure. The resulting optimal parameters are reported in the table. Systems with more than two activator binding sites are subject of overfitting and the corresponding values can be more or less degenerated. Up to three activator binding sites the retrieved parameters have relatively small excursion range, with higher binding sites number the fitting is totally degenerated and extrapolated parameters heavily cross compensate each other.

4.2 Systematic Study on the Effect of Multiple Activator Binding Sites in Promoters
Results

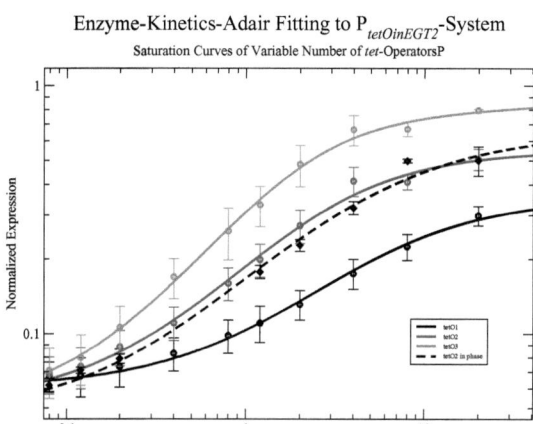

$P_{tetOinEGT2}$	S	B	K	c_1	c_2	$max_x(\varepsilon)$	$max_y(\varepsilon)$
$tetO1$	0.30	$6.06 \cdot 10^{-2}$	0.17			2.41	0.42
$tetO2$	0.25	$5.23 \cdot 10^{-2}$	0.34	1.41		0.89	0.56
$tetO3$	0.38	$4.80 \cdot 10^{-2}$	0.24	1.95	$1.07 \cdot 10^{-2}$	0.63	0.72
$tetO2$ in phase	0.30	$4.73 \cdot 10^{-2}$	0.27	0.68		0.95	0.54

Figure 34: $P_{tetOinEGT2}$ single dataset fitting to enzyme-kinetics-Adair equation (Eq.31). The fitting procedure is carried out by varying all binding parameters for each curve separately. The variable parameters are restricted to be greater than zero. K represents the apparent affinity of the rtTA to tet-operators (considered to be an association binding constant). c_i is the enhancement factor that modulates i^{th} binding. $tetO2$ in phase represents a construct in which the activator binding sites are spaced in a way to allow closer contact to adjacent rtTA bound to the DNA. $max_x(\varepsilon)$ and $max_y(\varepsilon)$ indicate the coordinates of the induction point of maximal sensitivity according to the corresponding fitted function. A basal expression term B added to the Adair equation, and a scaling factor S that multiplies it are additional free parameters used during the fitting procedure. The resulting optimal parameters are reported in the table. Systems with more than two activator binding sites are subject of overfitting and the corresponding values can be more or less degenerated. Up to three activator binding sites the retrieved parameters have relatively small excursion range, with higher binding sites number the fitting is totally degenerated and extrapolated parameters heavily cross compensate each other.

4.2 Systematic Study on the Effect of Multiple Activator Binding Sites in Promoters
Results

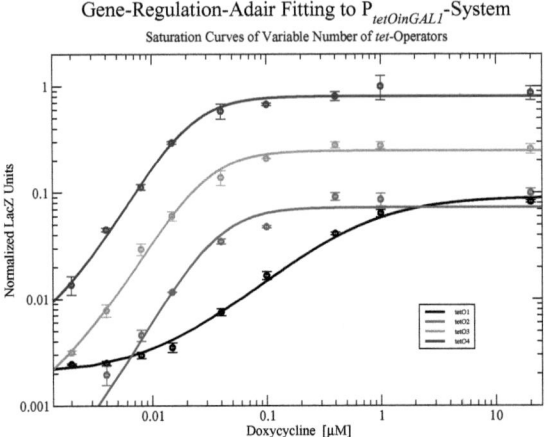

$P_{tetOinGAL1}$	S	B	K	c_1	c_2	c_3	$max_x(\varepsilon)$	$max_y(\varepsilon)$
tetO1	$8.96 \cdot 10^{-2}$	$1.93 \cdot 10^{-3}$	1.70				$8.54 \cdot 10^{-2}$	0.75
tetO2	$7.26 \cdot 10^{-2}$	$2.28 \cdot 10^{-8}$	0.79	$1.19 \cdot 10^{3}$			$7.71 \cdot 10^{-3}$	1.69
tetO3	0.25	$6.18 \cdot 10^{-4}$	1.02	386.87	$5.75 \cdot 10^{-8}$		$7.67 \cdot 10^{-3}$	1.56
tetO4	0.80	$3.47 \cdot 10^{-3}$	0.60	$1.05 \cdot 10^{3}$	$1.52 \cdot 10^{-6}$	$1.29 \cdot 10^{-6}$	$5.83 \cdot 10^{-3}$	1.61

Figure 35: $P_{tetOinGAL1}$ single dataset fitting to gene-regulation-Adair equation (Eq.38). The fitting procedure is carried out by varying all binding parameters for each curve separately. The variable parameters are restricted to be greater than zero. K represents the apparent affinity of the rtTA to tet-operators (considered to be an association binding constant). c_i is the enhancement factor that modulates i^{th} binding. $max_x(\varepsilon)$ and $max_y(\varepsilon)$ indicate the coordinates of the induction point of maximal sensitivity according to the corresponding fitted function. A basal expression term B added to the Adair equation, and a scaling factor S that multiplies it are additional free parameters used during the fitting procedure. The resulting optimal parameters are reported in the table. Systems with more than two activator binding sites are subject of overfitting and the corresponding values can be more or less degenerated. Up to three activator binding sites the retrieved parameters have relatively small excursion range, with higher binding sites number the fitting is totally degenerated and extrapolated parameters heavily cross compensate each other.

4.2 Systematic Study on the Effect of Multiple Activator Binding Sites in Promoters
Results

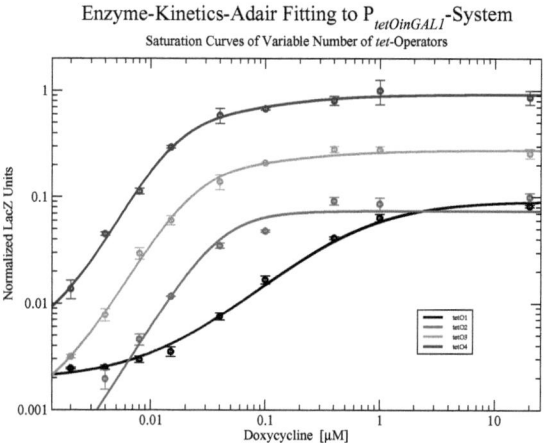

$P_{tetOinGAL1}$	S	B	K	c_1	c_2	c_3	$max_x(\varepsilon)$	$max_y(\varepsilon)$
tetO1	8.96·10^{-2}	1.93·10^{-3}	1.70				8.54·10^{-2}	0.75
tetO2	3.70·10^{-2}	4.14·10^{-8}	1.60	285.55			7.66·10^{-3}	1.67
tetO3	9.19·10^{-2}	6.20·10^{-4}	2.17	133.86	5.78·10^{-2}		6.52·10^{-3}	1.60
tetO4	0.23	3.48·10^{-3}	1.96	167.21	0.11	0.61	5.37·10^{-3}	1.66

Figure 36: $P_{tetOinGAL1}$ single dataset fitting to enzyme-kinetics-Adair equation (Eq.31). The fitting procedure is carried out by varying all binding parameters for each curve separately. The variable parameters are restricted to be greater than zero. K represents the apparent affinity of the rtTA to tet-operators (considered to be an association binding constant). c_i is the enhancement factor that modulates i^{th} binding. $max_x(\varepsilon)$ and $max_y(\varepsilon)$ indicate the coordinates of the induction point of maximal sensitivity according to the corresponding fitted function. A basal expression term B added to the Adair equation, and a scaling factor S that multiplies it are additional free parameters used during the fitting procedure. The resulting optimal parameters are reported in the table. Systems with more than two activator binding sites are subject of overfitting and the corresponding values can be more or less degenerated. Up to three activator binding sites the retrieved parameters have relatively small excursion range, with higher binding sites number the fitting is totally degenerated and extrapolated parameters heavily cross compensate each other.

4.2 Systematic Study on the Effect of Multiple Activator Binding Sites in Promoters
Results

In addition to the information provided by the fitted parameters obtained by Adair equation analysis, optimally fitting curves can be used to compute logarithmic sensitivity. Hence, information can be gained from Adair functions that accurately fit experimental datasets in order to analise systems' sensitivity independently on overfitting issues resulting from the fitting procedure itself. The Adair equations fitted to the studied experimental systems have inflection points and corresponding logarithmic sensitivities are represented by non-linear functions characterized by a maximum. The point of maximal sensitivity has been reported in the two last columns of the tables presented in these sections with the notation $max_x(\varepsilon)$ and $max_y(\varepsilon)$ indicating coordinates corresponding to induction point and extent of maximal logarithmic sensitivity, respectively. Besides being weakly influenced by partially populated datasets, maximal sensitivity points provide information about the range of system's activity corresponding to highest dynamicity. Logarithmic sensitivity analysis concords with the interpretation of Adair equation fitting and similarly suggests stronger influence of lower order binding events for promoter activation. Higher order binding events mainly contribute to scale absolute expression, not changing the profile of system's sensitivity functions but rather shifting them to lower induction range and thus playing minor role in system's dynamics.

A qualitative analysis of the equations inspired by the experimental findings has been used to shed light on the weight that different parameters exert on the overall behavior of saturation curves. In the systems studied in these sections, the contribution of binding sites after the second/third seems to have negligible consequences for cooperativity. The main contribution these additional binding sites bring to the system consists of shifting the sensitivity range of the different constructs to lower induction ranges. The mechanism underlying this effect is however, as outlined by theoretical curves generated by the Adair equations presented in Fig.37, of statistical nature and independent on binding cooperativity. The apparent binding affinity of promoters with higher activator binding sites increases simply because activators are statistically advantaged to bind promoters with multiple binding sites compared to single binding site ones (see Eq.28). The role of cooperativity for higher binding terms has almost no influence on the qualitative features of saturation curves within the experimentally accessible range and by setting different enhancement factor values one cannot distinguish different saturation curves. In contrast, the statistical term associated to multiple binding site systems displays a clear pattern characterized by displacing saturation curves to lower induction leaving their qualitative properties largely unchanged ($max_x(\varepsilon)$ changes, $max_y(\varepsilon)$ remains constant).

These results show that non-linearity in promoter activation mainly rely on low order binding events which are responsible for large changes in promoter activation profiles. Higher order binding contributes instead to shift promoter activity to lower range of induction in a more linear manner.

4.2 Systematic Study on the Effect of Multiple Activator Binding Sites in Promoters
Results

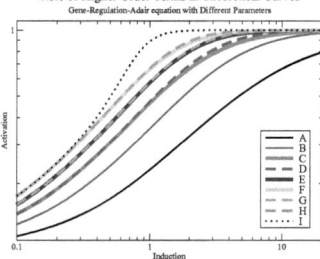
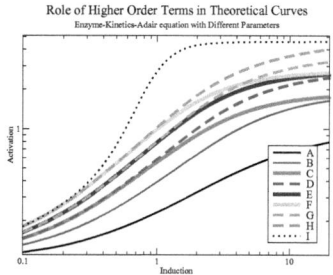

Theoretical Curves						GenReg		EnzKin	
	n	K	c_1	c_2	c_3	c_4	$max_x(\varepsilon)$	$max_y(\varepsilon)$	$max_x(\varepsilon)$	$max_y(\varepsilon)$
A	1	0.17					1.86	0.52	1.86	0.52
B	2	0.17	2				1.03	0.59	1.34	0.70
C	3	0.17	2	0			0.68	0.61	0.85	0.74
D	3	0.17	2	1.29			0.73	0.62	1.09	0.82
E	4	0.17	2	1.29	0		0.55	0.64	0.81	0.88
F	5	0.17	2	1.29	0	0	0.44	0.65	0.64	0.91
G	4	0.17	2	1.29	1		0.56	0.64	0.91	0.91
H	5	0.17	2	1.29	1	1	0.46	0.65	0.77	0.97
I	5	0.17	2	1.29	10	10	0.58	1.00	0.57	2.18

Figure 37: Theoretical curves inspired from $P_{tetOinEGT2}$ system and represented with gene-regulation-Adair equation (Eq.38, left plot) and enzyme-kinetics-Adair equation (Eq.31, right plot). As suggested from experiments, the role of cooperative enhancement for binding events higher then third order exerts little contribution to promoter activation. This is shown in curves E and G, as well as in F and H, where complete absence of enhancement or simple independent binding cannot be clearly distinguished especially under the assumption of regulation according to gene-regulation-Adair equation. Even for curve D, in which according to experiments and sequential fitting to gene-regulation-Adair equation (see Fig.27), a positive enhancement is present, the situation cannot be clearly distinguished from the case of complete absence of enhancement (curve C). In contrast to weak, hardly detectable enhancement factors, curve I shows a case of stronger cooperativity for third and fourth binding. The last four columns of the table report induction point of maximal logarithmic sensitivity computed using gene-regulation-Adair (left) and enzyme-kinetics-Adair equation (right).

4.2 Systematic Study on the Effect of Multiple Activator Binding Sites in Promoters
Results

4.2.2.4 Discussion

The approach presented in this section has been inspired by experiments performed in an earlier stage of the present work where some relevant aspects have been considered in relation to the role of multiple activator binding sites in promoter regions and the corresponding mode of regulation in their biological context. A methodic study of promoters regulated by variable number of activator binding sites has thereby been carried out using synthetic components.

Activator binding sites of three different promoter systems (promoter sequences of *CYC1*, *EGT2* and *GAL1* genes) have been replaced by repetitive *tet*-operator blocks and subsequently exposed to doxycycline induction gradients. The effect of different number of binding sites, the role of different core promoters as well as distinct activator binding configuration have been systematically investigated. The outcome of the experiments pointed out that the three analyzed gene systems integrate a common stimulus mediated by same components displaying significantly different activation modes. Comparing non-linear fitting procedures of experimental datasets interpreted using Hill and Adair functions together with the use of logarithmic sensitivity analysis revealed unexpected role of promoter sequences in integrating same stimuli differently and reminded the mechanisms described in the section 'Context-Dependent Effects on Gene Regulation'. Promoter activation patterns observed in this section highlighted the capacity of promoters to change their activation profile based on the number of activator binding sites regulating their activity and, despite the fact that this behavior is largely determined by core promoter sequences, the importance of low order binding events emerges as a common feature among all studied systems. The addition of a second binding site applies in fact more drastic changes to promoter's receptivity and reactivity with respect to higher order binding events. This highlights the importance of the non-linear character of low order binding. Higher order binding contributes indeed mainly to modulate sensitivity ranges in a more linear manner without strongly affecting the qualitative features of saturation functions.

4.2 Systematic Study on the Effect of Multiple Activator Binding Sites in Promoters
Results

4.2.3 Semi-Synthetic Systems

> **Motivation:** GAL1 promoter shows interesting activation patterns and displays high cooperativity within the full-synthetic experimental framework. Gene specific elements recruited by the native activator (Gal4p) are supposed to have an influence on the natural process of GAL1 expression and their role can be studied by replacing the gene encoding for Gal4p by a synthetic, controllable gene module.
> **Aims:** Externally modulate expression of Gal4p in order to gradually control GAL1 promoter activation mediated by the natural components participating to the reaction under natural conditions, thereby taking into consideration gene specific effects.
> **Results:** Non-linear effects dominate the semi-synthetic system displaying switch-like activation pattern that could have its basis at the level of transcriptional machinery organization as a consequence of promoter specific effects.

Among the promoters studied with the full synthetic experimental approach, the most interesting from the point of view of cooperative activator binding was represented by the $GAL1$ promoter (P_{GAL1}). This promoter has been indeed studied extensively and important information concerning its regulation is available [110, 7, 26, 25, 27, 9, 6, 28, 29]. Moreover, binding data of interactions involving regulators of the GAL network with their target binding sites are known from *in vitro* experiments [109, 111].

In order to get more insights in the regulation of P_{GAL1}, we extended the study carried out in the full-synthetic context to an experimental setup more closely representative of the natural regulatory process. Instead of the synthetic binding partners (rtTA–*tet*-operators) we used the native ones (Gal4p-$Gal4pBS$). The approach consists in combining a synthetic circuit driving activator's expression together with a natural component represented by the native promoter sequence, which later has been subject to genetic modifications. More precisely, in the resulting semi-synthetic systems we dissected the activator-promoter (Gal4p-P_{GAL1}) interaction into a synthetic, externally tunable module, which replaces the natural $GAL4$ gene driving Gal4p expression. The transcriptional activator Gal4p can therefore interact with the natural gene regulatory sequence of the $GAL1$ gene in a externally controlled manner. Here we use tetR-Ssn6 (transcription factor consisting of tetR DNA binding domain, Ssn6 repressor domain), which binds to two *tet*-operators placed in the regulatory region of the $GAL4$ gene ($P_{tetO2inGAL4}$). In absence of doxycycline, tetR-Ssn6 represses the expression of $GAL4$. With addition of doxycycline the repression is relieved and $GAL4$ is expressed in a graded way proportional to external induction. Increasing doxycycline increases the level of the activator (Gal4p) which in turn binds to the target regulatory sequence of $GAL1$, that finally drives the expression of a reporter gene (see Fig.38). Finally, the constitution of the natural P_{GAL1} has been modified in the number and affinity of the binding sites and the resulting activation functions compared with respect to the wild type sequence (see Fig.39).

4.2 Systematic Study on the Effect of Multiple Activator Binding Sites in Promoters
Results

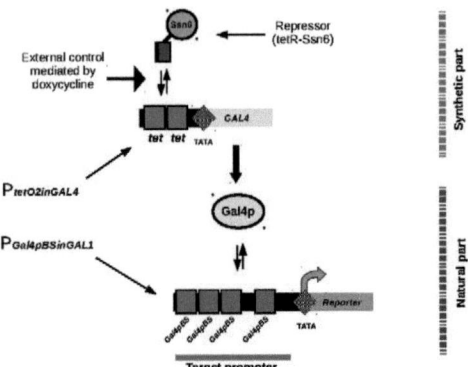

Figure 38: Principle of semi-synthetic system to study the effect of number and type of Gal4p binding sites ($Gal4pBS$) on gene activation. The synthetic part can be controlled externally by adjusting doxycycline concentration and is responsible for the production of Gal4p, the transcriptional activator that binds the target promoter under study (natural part of the system). The structure of the target promoter (P_{GAL1}) is maintained constant except for the four Gal4p binding site sequences. Both number and sequences of binding sites have been modified to study their effect on promoter activation.

4.2.3.1 System Description

In order to study the pure effect of Gal4p binding to it's consensus sequence, we carried out this part of the work in yeast strains devoid of the GAL genes repressor (Gal80p) and in glucose-free growing conditions (2% raffinose and 0.5% galactose). Gal80p is normally responsible for repressing GAL genes in absence of galactose through inhibition of Gal4p transcriptional activation and glucose-free conditions prevent GAL genes to be negatively regulated by glucose repression mechanisms.

The system studied in this section has been denominated $P_{Gal4pBSinGAL1}$ where $Gal4pBS$ stands for the binding site configuration of a given construct. For instance, the Gal4p activator binds different target binding sites in the yeast genome characterized by a consensus sequence ($CGG[N]_{11}CCG$, N representing unspecified nucleotides). The advantage of studying such consensus is twofold: It allows to maintain binding specificity and at the same time can display variable range of affinities for Gal4p. The binding sites used in this part of the work are the four different ones present in the native P_{GAL1} organization (named $GAL1BS1$ - $GAL1BS4$) and the Gal4p binding site present in the promoter of the $GAL80$ gene, $GAL80BS$ (see Appendix I for sequences). Type and number of binding sites have been varied in the course of this section and the promoter is designated with the notation $[Gal4pBS]X$, $Gal4pBS$ indicating the operator type and X its copy number in a given construct (see Fig.39).

With the semi-synthetic approach we have principally studied the effect of same induction conditions acting on different target promoters. The principal difference that

4.2 Systematic Study on the Effect of Multiple Activator Binding Sites in Promoters
Results

we focused on has been the number of activator binding sites present in the target promoters. In addition, the affinity of single binding site sequences was also varied.

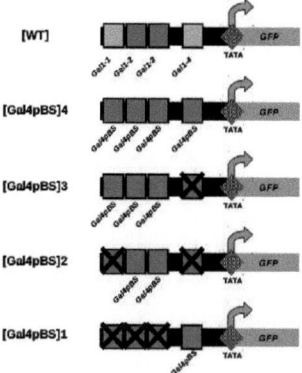

Figure 39: Scheme representing the modifications applied to the different target promoters (P_{GAL1} constructs containing different number of binding sites for Gal4p (denoted with $Gal4pBS$)) and compared with the wild type ($[WT]$) promoter architecture in which the two central binding sites have stronger activator affinity than the flanking ones (see Fig.42).

4.2.3.2 Results

A switch-like response is obtained by doxycycline induction experiments and the saturation curves resulting from this approach indicate that different constructs have different maximal expression values, moreover, this upper bound of promoter activation is not only determined by operator number, but also by their affinity for Gal4p (see Fig.40). Interestingly, all curves reach different V_{max} values at nearly the same induction level, consequently they all have about the same K value. Non-linearities in the molecular cascade connecting the synthetic module to the natural part of the semi-synthetic system could in effect be the cause of the observed effect. It can be argued for example that at high induction of the synthetic $GAL4$ gene, Gal4p accumulates in the cytoplasm because nuclear transport reaches saturation representing a bottleneck in the dynamics of the system. Accordingly, after a certain induction level, more activator is produced by the synthetic module but activator's nuclear concentration remains fixed at a maximal level independent on external induction, so that no further activation of the natural module can be attained. Alternatively, the source of non-linearity could lie in an effect driven by specific elements recruited by Gal4p. Similarly to the phenomenon observed in the section 'Context-Dependent Effects on Gene Regulation', specific components participating to the process of the native $GAL1$ promoter activa-

4.2 Systematic Study on the Effect of Multiple Activator Binding Sites in Promoters
Results

tion could promote changes in its receptiveness and thereby explain the high sensitivity displayed by all studied promoter versions. Hence, cooperativity at the level of the transcriptional machinery organization could explain the experimental findings consistent with same apparent binding constant and different maximal expression values.

Comparing the effect of high $GAL4$ expression induced by the semi-synthetic module (see Fig.40), and induced by simple exposure of wild type cells to saturating galactose concentration (see Fig.41), similar activation levels are reached by the different constructs. The affinity of the binding sites used in the target promoter of the semi-synthetic system reflects the situation represented by the native system induced by the natural stimulus. For example, the two individual central binding sites of the native P_{GAL1} have stronger affinity for Gal4p than $GAL80BS$ (compare $GAL1BS2$, $GAL1BS3$ and $GAL80BS$ in Fig.42). This difference in individual binding affinity is present in both the semi-synthetic system and the native system induced by its natural stimulus (compare $[GAL1BS2-GAL1BS3]$ with $[GAL80BS]2$ in Fig.41). Similarly, comparing the two constructs containing four activator binding sites, the observed differences are also in accordance with the individually measured contributions of each binding sites taken separately (see Fig.42). In this case the wild type sequence ($[WT]$) contains two strong and two weak binding sites and consequently has lower level of expression at maximal activation compared to $[GAL80BS]4$ which contains four strong binding sites. This implies that in both cases, the externally tunable semi-synthetic system induced by high activation level of the synthetic module as well as under natural conditions induced by high galactose concentration, the system reaches saturation in a similar way and that the attained saturation level is dependent on number and affinity of Gal4p binding sites composing the target promoter.

4.2.3.3 Discussion

Information about P_{GAL1} activation could be obtained from the semi-synthetic approach. Comparing with a simpler approach consisting of measuring promoter maximal activation by applying galactose induction to wild type cells, the semi-synthetic approach has the advantage to give access to saturation functions instead single expression data points. An open question remains to be clarified concerning the source of non-linearities responsible for system's behavior. The fact that the fitted Hill coefficients have very high values, and moreover this also applies to the construct containing a single Gal4p binding site, indicates that an unknown effect is acting at the level of the connection between synthetic module and target promoters playing a role in non-linearizing system's response. This effect is responsible for the observed switch-like character of the system and is potentially relying on the transcriptional machinery organization as a consequence of gene specific effects. We consequently decided to further reduce the level of invasivity in order to study P_{GAL1} activation in a context as similar as possible to its native one, anyway conserving the same aim to experimentally retrieve saturation curves instead single expression data points. So we directed our efforts to develop the third approach, called natural system approach, presented in this part of the work.

4.2 Systematic Study on the Effect of Multiple Activator Binding Sites in Promoters
Results

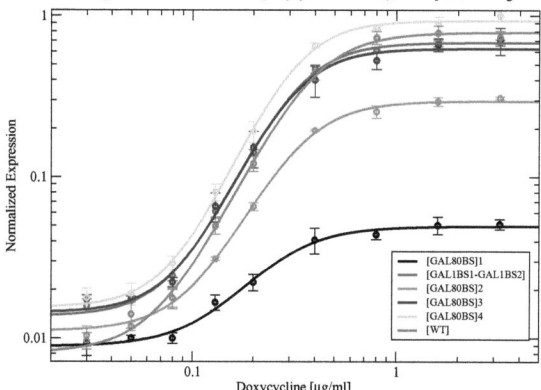

$P_{Gal4pBSinGAL1}$	B	V_{max}	K_D	n_H
$[GAL80BS]1$	$8.88 \cdot 10^{-3}$	$4.06 \cdot 10^{-2}$	0.26	2.51
$[GAL1BS2-GAL1BS3]$	$1.37 \cdot 10^{-2}$	0.67	0.32	2.88
$[GAL80BS]2$	$1.11 \cdot 10^{-2}$	0.29	0.33	2.75
$[GAL80BS]3$	$1.44 \cdot 10^{-2}$	0.61	0.31	2.95
$[GAL80BS]4$	$1.55 \cdot 10^{-2}$	0.92	0.32	2.99
$[WT]$	$8.14 \cdot 10^{-3}$	0.78	0.38	2.71

Figure 40: Synthetic module regulates the expression of the *GAL4* gene through tetR-Ssn6 which is recruited inversely proportional to the applied doxycycline concentration, the expression of the activator Gal4p is then directly proportional to doxycycline concentration. The activation of the target promoter of Gal4p, P_{GAL1}, is shown as a function of its composition: Number and strength of Gal4p binding sites represent the difference between target promoters. Error bars are calculated with standard error of the mean from three experiments and expression values are normalized to their maximal value. The datasets were fitted with non-linear regression to Hill equation (Eq.47) to which a basal expression term (B, corresponding to the expression value measured at zero doxycycline) has been added and kept fixed during the fitting procedure. The resulting optimal parameters are reported in the table.

4.2 Systematic Study on the Effect of Multiple Activator Binding Sites in Promoters
Results

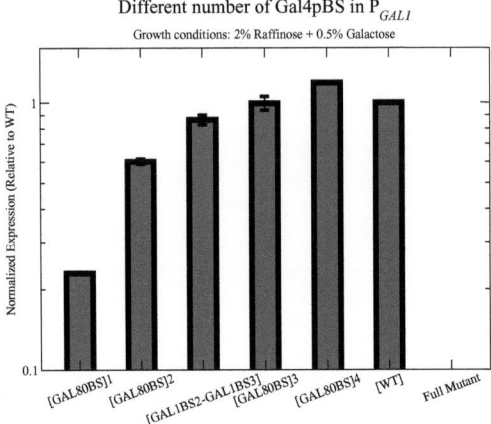

Figure 41: Variable number of Gal4p binding sites in P_{GAL1} under natural conditions. The constructs used in Fig.40 have been transformed into wild type strains (W303A) and induced (0.5% galactose) in glucose-free conditions to avoid glucose repression. GFP fluorescence was red out by flow cytometry. Expression corresponding to the full mutant construct was lower than the background strain containing an empty plasmid.

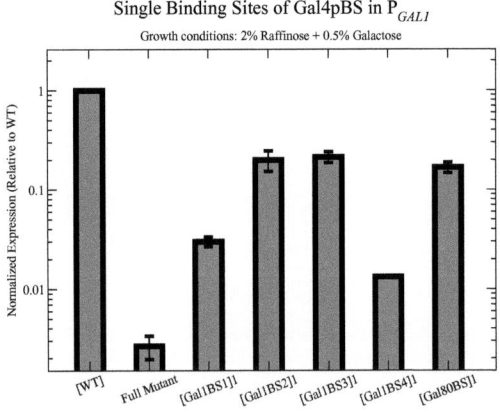

Figure 42: Expression of single Gal4p binding sites in P_{GAL1} under natural conditions. The different constructs have been transformed into wild type strains (W303A) and induced (0.5% galactose) in glucose-free conditions to avoid glucose repression. GFP fluorescence was red out by flow cytometry.

4.2.4 Natural Systems

> **Motivation:** Previous work pointed out the role of gene specific effects as potential cause of non-linearities observed in gene activation profiles. In order to account for these effects and get further insights in the gene activation process, a non invasive technique minimally perturbing cellular physiology is needed. A genetic approach based on synthetic promoter libraries of variable affinity for selected transcriptional regulators can be used to access saturation curves reflecting activation of genes of interest. Promoters containing multiple activator binding sites can thereby be represented in terms of activation gain with respect to their single binding site contributions, thereby allowing comparison between different analyzed systems.
> **Aims:** Gradually induce gene systems through activator binding site affinity variation thereby maintaining the concentration of the selected transcriptional regulator under study unperturbed. Since regulator concentration and affinity compensate each other and are treated as equivalent variables in reaction kinetics theory, it can be shown that the same information obtained by inducing promoters of constant affinity with a gradual activator concentration increase can also be retrieved by maintaining activator concentration unchanged and applying graded variation in its affinity.
> **Results:** The methodology has been firstly validated on a synthetic test system allowing to control several determining aspects critical for the experimental verification of theoretical concepts. Subsequently, the approach has been applied to the GAL1 promoter and a saturation curve could be obtained over broad range of induction. GAL1 promoter displayed high reactivity at low induction but graded response when exposed to higher induction levels. The role of multiple binding sites contributed to a more-than-additive increase of expression compared to the promoter activated by a single binding site revealing that activators can cooperate not by direct interaction but through components of the transcriptional machinery.

We refer to 'natural systems' when we make use of promoter libraries of variable affinity for a certain native transcription factor. We call this method 'natural' for the simple reason that among the other ones presented previously, this is the least invasive and thus it would be the method of choice to get insights in the natural process of gene regulation with minimal perturbation affecting cellular physiology.

This technique involves the simple introduction in cells of an additional copy of the promoter sequence of interest leaving the native one untouched and thereby avoiding the effect of propagating perturbations to the cellular machinery. While the endogenous copy of the gene under study executes its function unperturbed, the additionally introduced copy can drive in parallel the expression of a reporter gene. This additional copy is used to access information about the activation state of the promoter of interest and, in order to obtain a saturation function that characterizes the activation of that promoter (and not single data points), a series of identical promoters differing *only* in the affinity of the activator binding site sequence is introduced in different cells of the same yeast strain. The promoter library of variable affinity is obtained by introducing gradual mutations in the consensus binding site sequence for a given transcription factor and a reporter gene allows to read out the promoter activation level. So, by exposing cells that contain the promoter library to the same conditions, one readily accesses the same information obtained by varying the concentration of regulator in the nucleus of the cell. This is actually supported by the mathematical formalism developed for the description of chemical binding processes where the concentration of a ligand and its binding affinity are considered equivalent variables. Reaction kinetics equations assume in fact that

4.2 Systematic Study on the Effect of Multiple Activator Binding Sites in Promoters
Results

both activator concentration (x) and affinity (K) compensate each other. Consequently, these two variables appear in the same form in reaction kinetics equations. It is actually considered that whenever an activator binding reaction occurs with low probability (due to the weakness of its binding affinity K), it can be compensated by the presence of high activator concentration: The probability that rare events occur is compensated by multiplying those single events. For the same reason, if low amount of interacting molecular partners is limiting the probability of a reaction to occur, high binding affinity can have a compensatory effect by making rare molecular encountering highly probable to result in a successful reaction outcome.

In order to obtain information on cooperativity, and, more generally, to be able to observe and quantify the contribution of multiple activator binding sites in promoter sequences, the methodology presented here is decomposed in two parts, called *calibration* and *measurement*. The goal of the calibration phase is to determine individual contributions that single binding site sequences bring to system's activation. The sequences used for calibration are consensus binding site sequences specific for the studied activator. Applying different mutations to consensus sequences, the binding affinity for the studied activator can be modulated yet conserving specificity. Once the system calibrated, thus knowing the extent of system activation that every individual mutant binding site in the library contributes to, the measurement phase consists of introducing for each mutant sequence used in the single operator library, its corresponding construct composed of the same affinity mutant present in multiple copies. Comparing each pair of single/multiple binding sites using a set of sequences expressing graded activity (operator library), one can obtain functions that describe the gain in promoter activation realized upon the presence of multiple binding sites in the regulatory sequence under study (see Fig.43).

Varying transcription factor concentration *in vivo* is for obvious reasons not feasible in most of the cases and whenever possible it often implies important consequences for cellular physiology. Thus, the method of promoter libraries of variable affinity is a potential solution to characterize promoter saturation functions in a minimally invasive way that reduces to the expression of a reporter gene.

4.2.4.1 System Description

The natural approach has been initially tested to study rtTA–*tet*-operators interaction in the *EGT2* core promoter (with same experimental setup described in the section 'Full-Synthetic Systems'), and only later applied to the Gal4p-*Gal4pBS* binding partners of the *GAL1* promoter (P_{GAL1}).

Representing an exception among transcriptional factors, doxycycline tunable rtTA–*tet*-operators interaction offers a unique opportunity to externally fix a certain amount of active regulator in living cells and consequently allows to study the effect of binding site mutations under different activator concentrations. Testing the natural system approach on rtTA–*tet*-operators interaction allowed us to ascertain the validity of the technique on a system that has been previously investigated in this study on one hand, and which is well characterized by mutational studies found in the literature on the other hand. However, the major advantage of this system consists of the possibility to access saturation curves retrieved with the natural approach, and to compare the outcome with the

4.2 Systematic Study on the Effect of Multiple Activator Binding Sites in Promoters
Results

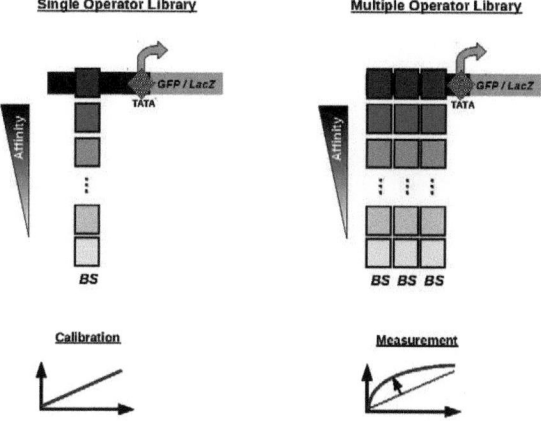

Figure 43: Promoter libraries of variable affinity. The natural system method consists in creating two libraries containing mutant sequences of variable affinity for a given regulator. The single operator library is realized by applying mutations in the activator consensus sequence that affect promoter activation in a graded way. The multiple operator library uses the same mutants as in the single operator library, with the difference of being present in blocks of repetitive units within the studied promoter. Calibration of the system is achieved by using the single operator library while the measurement phase is carried out using the multiple operator library: Under same external conditions, the deviation in promoter activation between single and multiple operator libraries reflects the gain represented by the presence of multiple activator binding.

4.2 Systematic Study on the Effect of Multiple Activator Binding Sites in Promoters
Results

corresponding activation profile obtained by doxycycline driven induction (similarly to the methodology applied to the full-synthetic approach).

After having assessed validity of the experimental design on the rtTA–*tet*-operator test system, we could further extend the study by applying the natural system approach to the *GAL1* promoter and consequently access information concerning Gal4p-*Gal4pBS* binding under natural conditions.

4.2.4.2 Testing the Natural Systems Approach: rtTA–*tet*-operators Interaction

4.2.4.2.1 Calibration: Single *tet*-operator Library

The *tet*-operator (*tetO1* type) is a 19 bp palindromic sequence to which the rtTA homodimer binds. The advantage of such properties for our study relies on the symmetry of these binding partners at the molecular level. The binding site sequence can be altered by single nucleotides mutations in one half of the operator only, resulting in conserved specificity on one side and altering affinity for rtTA on the other side. Mutagenesis studies have been performed in *Escherichia coli* and the effect of every nucleotide mutation on TetR binding has been reported in [23]. Based on this information we synthetically obtained different single mutant operator sequences that were inserted in the core promoter of the *EGT2* gene (according to the procedure presented in the section 'Full-Synthetic Systems'), transformed into wild type *Saccharomyces cerevisiae* (W303A), mated with an analogous wild type strain (W303α) containing P_{CLN3} driven rtTA, and expression of diploids was measured by flow cytometry under high inducing conditions (20 µM doxycycline). Despite the fact that in yeast the basal activity due to the *EGT2* core promoter is roughly 2-fold higher than in the bacterial system (8% in *Escherichia coli* ([23]) and 17% in yeast with respect to the wild type), the outcome showed that both organisms expressed highly correlated activities (correlation coefficient > 0.99) in relation to the same mutations (see Fig.44).

4.2 Systematic Study on the Effect of Multiple Activator Binding Sites in Promoters
Results

```
-9 -8 -7 -6 -5 -4 -3 -2 -1  0  1  2  3  4  5  6  7  8  9
 A  C  T  C  T  A  T  C  A  T  T  G  A  T  A  G  A  G  T
```

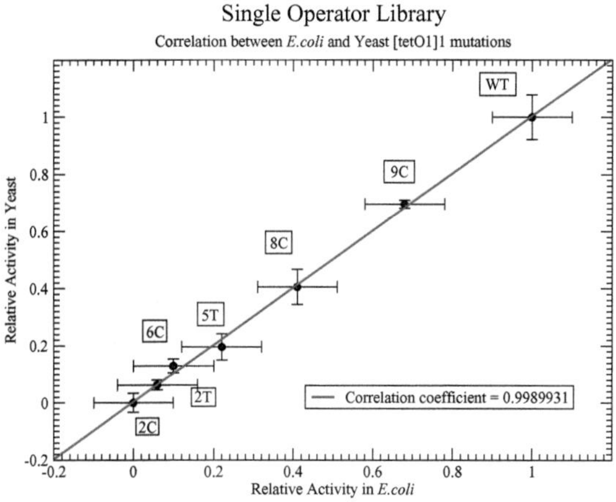

Single Operator Library
Correlation between *E.coli* and Yeast [tetO1]1 mutations

Figure 44: Calibration phase of single operator library. Correlation of single operator library measured in *Escherichia coli* (according to expression values in [23]) and in *Saccharomyces cerevisiae* at high inducing conditions (20 µM doxycycline). The upper part illustrates the palindromic *tet*-operator sequence (*tetO1* type) with positions indicated with positive numbers referring to one half and negative numbers referring to the other half of the sequence. The lower plot shows the effect of single nucleotide mutations indicated by boxes, where numbers designate the position according to nucleotide coordinates described in the upper sequence, and letters correspond to substitutive nucleotides at that position. The basal expression value is 8% and 17% of the wild type sequence expression in *Escherichia coli* and *Saccharomyces cerevisiae*, respectively. The correlation coefficient is > 0.99, reproducibility of LacZ expression measurements in bacteria lie within ±10% ([23]) and error bars of yeast expression measurements show standard deviations from at least 3 replicates.

4.2.4.2.2 Measurement: Multiple *tet*-operator Library

Once assessed how single *tet*-operator mutants influence gene activation individually, the next step consists of testing the same mutant sequences in corresponding promoter systems containing multiples of their copies. Hence, we synthetically obtained multiple operator repeats consisting of four identical *tet*-operator sequences (*tetO1* type, see Appendix I for sequences) spaced by 6 or 8 bp that we introduced in the core promoter of the *EGT2* gene (according to the procedure presented in the section 'Full-Synthetic Systems'). The resulting vectors were subsequently transformed into wild type strains of *Saccharomyces cerevisiae* (W303α) containing P_{CLN3} driven rtTA. Similarly, the corresponding single *tet*-operator library has also been transformed into the wild type W303α strains containing P_{CLN3} driven rtTA and both resulting series of haploid strains have been measured at the same time by flow cytometry at constant doxycycline concentrations (20 µM or 6 µM).

When expressing multiple and single operator systems as a function of the same input it is convenient to combine the two corresponding outputs and represent the deviations between each other in a parametric manner (see section 'Parametric Representation of Gene Expression Data'). Experimentally, the current setup involves two promoter systems containing different number of the same binding site sequences for a common activator. It is thereby possible to combine single and multiple *tet*-operator libraries into a parametric representation considering the variable affinity of mutant operators as common input. Afterwards, measuring deviations between the obtained graded expression curves allows to quantify the influence of multiple with respect to single activator binding sites in promoter sequences at different ranges of induction.

Moreover, the use of synthetic components offers the opportunity to externally tune the active portion of rtTA in cells and thereby the possibility to carry out measurements of the same system under different conditions determined by the activator concentration. The setup of this test system allows to emulate experimental situations where this same approach is applied to natural systems with different activator concentration. According to theory (see section 'Parametric Representation of Gene Expression Data'), the obtained parametric curves should return unbiased information about system's parametrization avoiding that different activator concentrations influence measures inherent system's properties. The result of exposing cells containing libraries of mutant *tet*-operators resulted indeed in the expected distortion of the calibration curve and in addition, under different induction conditions (with different activator concentrations), the system responded according to expectations as well by displacing experimentally accessible regions of the same curve to different expression ranges (see Fig.45 and compare with theoretical expectations (Fig.12)).

4.2 Systematic Study on the Effect of Multiple Activator Binding Sites in Promoters
Results

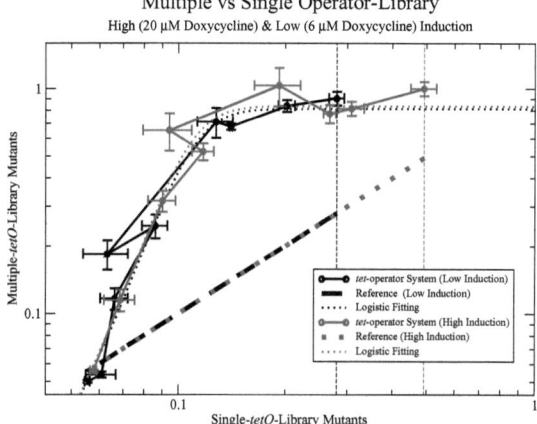

Figure 45: Measurement phase of multiple operator library. The x-axis represents the range of expression covered by the single operator library while on the y-axis by the multiple operator library (consisting of a block of four *tet*-operators). Measurement curves of the multiple operator library corresponding to high (red lines, 20 µM doxycycline) and low (black lines, 6 µM doxycycline) induction are obtained by combining expression values of x and y-coordinates while reference systems are constructed by the same procedure but comparing single *versus* single operator library, thus giving a diagonal line (thick, dashed and dotted lines). Thin dashed lines indicate the range of activity of single operator library at high (red) and low (black) induction (compare with Fig.12). Error bars are calculated with standard error of the mean from three experiments and expression values are normalized to their maximal value. Thin dotted lines are fitted functions to the datasets (standard logistic function (Eq.59) where the resulting parameters are: $A0 = 1.59 \cdot 10^{-3}$; $A1 = 1.96 \cdot 10^{-3}$; $A2 = 63.28$; $A3 = 2.60 \cdot 10^{-9}$ for the low induction and $A0 = 1.38 \cdot 10^{-3}$; $A1 = 1.65 \cdot 10^{-3}$; $A2 = 66.31$; $A3 = 5.40 \cdot 10^{-11}$ for the high induction curve. The experiments were obtained with strains and plasmids detailed in Tables 1, 2, 3.

4.2 Systematic Study on the Effect of Multiple Activator Binding Sites in Promoters
Results

4.2.4.2.3 Proof of Concept

The mutant *tet*-operator library could further be used to check the same principle described above in an alternative way. Since activator's affinity K and concentration x can be considered as dependent variables (based on the mathematical formalism used to describe reaction kinetics (see section 'Data Analysis & Mathematical Tools')), the validity of the methodology presented above where activator's binding site affinity has been varied and activator kept at constant concentrations, can also be proved equivalently by varying activator concentration and fixing affinity to constant values.

Thus, because of the advantage of using a synthetic test system such as rtTA–*tet*-operator, same material and principle used to check the validity of the methodology presented earlier could be used to carry out an additional validation check of the presented approach by simply swapping the equivalent variables x and K. Previously, a series of affinity mutants (variable K) were exposed to constant induction conditions (two different x values that was kept constant during the experiment). Subsequently, two members of the promoter libraries (constant K) were chosen to be exposed to doxycycline gradients. The two pairs of mutants (from single and multiple operator libraries), WT and $5T$, were gradually induced by doxycycline and correlated through parametric representation resulting in the curves shown in Fig.46 together with the dataset shown in Fig.45. The datasets are compared by normalizing single operator constructs to one: Highest induction or affinity mutant of the single operator library (represented in both cases by the wild type *tet*-operator at 20 µM doxycycline) was set to reach the x-axis value of one, single operator mutants exposed to lower induction or of lower affinity were thereby scaled proportionally and reached lower expression values. Multiple operators were also scaled accordingly and could thereby reach expression values higher than one. Experiments show that both ways of representing the rtTA–*tet*-operator system lead to overlapping curves and that, according to theory, the same information obtained by varying activator concentration using a constant affinity receptor can be equivalently accessed by fixing activator level to a constant concentration and gradually varying the affinity. The outcome is thus showing that promoter libraries of variable affinity can be efficiently used to study gene activation according to the natural approach methodology.

4.2 Systematic Study on the Effect of Multiple Activator Binding Sites in Promoters
Results

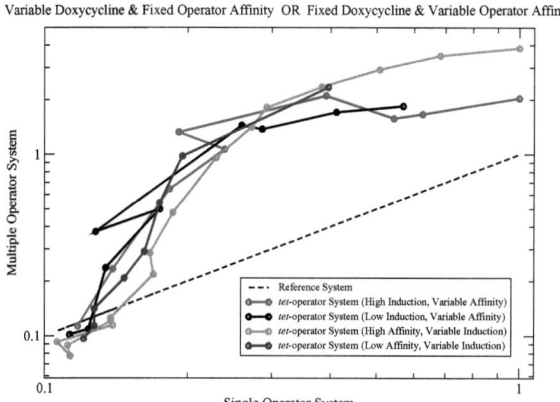

Figure 46: Equivalent variable compensation in rtTA–*tet*-operator system. Two sets of curves generated by the same system are compared. In the first set (same set presented in Fig.45), operator affinity was varied gradually at two different activator concentration levels: Single and multiple operator libraries were exposed to high (20 µM doxycycline) and low (6 µM doxycycline) induction (red and black curves, respectively) and their respective expression correlated parametrically using binding site affinity as common parameter. In the second set, operator affinity was kept constant at two different levels and the measurement carried out under graded doxycycline induction: Single and multiple operator constructs containing WT (high affinity, wild type (100% activity)) and $5T$ (low affinity, 22% of wild type activity) operators (green and blue curves, respectively) were exposed to the same doxycycline gradient and correlated parametrically using doxycycline concentration as common parameter. To combine the two different sets the expression of wild type single operator at 20 µM doxycycline was used as reference.

4.2 Systematic Study on the Effect of Multiple Activator Binding Sites in Promoters
Results

4.2.4.3 Application of the Natural System Approach to *GAL1* Promoter Activation

4.2.4.3.1 Calibration: Single *Gal4pBS* Library

Instead of performing a systematic mutagenesis of the activator binding site consensus, as it has been done for the *tet*-operator library, the calibration phase for measuring the affinity of Gal4p-*Gal4pBS* system was more suited to be approached differently. Since Gal4p activator binds several target genes in the genome of *Saccharomyces cerevisiae*, each differing in number and sequence of the relative binding sites, the many operator variants known to specifically bind Gal4p have been individually analyzed in order to identify candidates best suited for being employed in the measurement phase of the study. Dealing with *GAL1* promoter system required thereby to perform a screen over all native binding sites present in the yeast genome that were proved to specifically bind Gal4p. The activity of each of these binding sites was studied by cloning the different variants of the Gal4p consensus sequence ($CGG[N]_{11}CCG$) in the fourth position of P_{GAL1} (according to the description of $[Gal4pBS]1$ shown in Fig.39) and measuring the corresponding activities under high inducing conditions (0.5% galactose added to a 2% raffinose background). The screening outcome represented in Fig.47 revealed that the vast majority of native binding site sequences have intermediate-strong affinity for Gal4p and that an important decrease in affinity could be attained by mutating the first nucleotide in the consensus sequence recognized by Gal4p. Interestingly, these results revealed in a first time that native sequences of Gal4p binding sites display relatively narrow range of affinities. Indeed, the single operators were ranging roughly 100-fold in expression with respect to the fully mutated promoter (indicated as 'Full Mutant' in Fig.47, in which all binding sites were mutated in their consensus sequence thus abolishing Gal4p binding ($CGG[N]_{11}CCG \rightarrow GCC[N]_{11}GGC$)) but the expression of all of the natural binding sites was situated between ten and hundred times full mutant expression. In a second time the screening has been extended to more candidates in order to reach lower binding site strength using artificial operator sequences displaying expression levels closer to the one of the full mutant. Consequently, a single operator library of logarithmically graded expression constructs could be constituted and employed for calibrating the system.

4.2 Systematic Study on the Effect of Multiple Activator Binding Sites in Promoters
Results

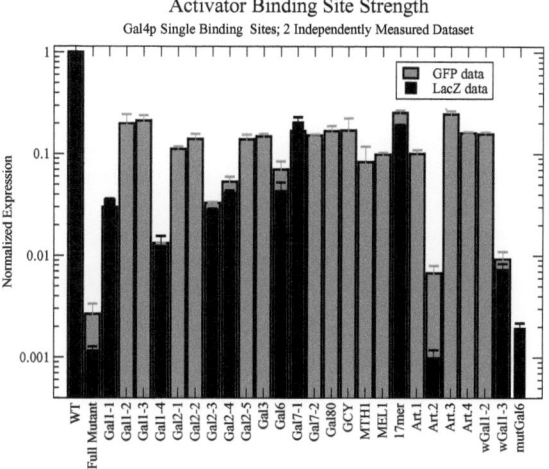

Figure 47: Gal4p binding sites affinity. Different variants of the Gal4p consensus sequence ($CGG[N]_{11}CCG$) were cloned in the fourth position of P_{GAL1} (according to the description of $[Gal4pBS]1$ shown in Fig.39) and the corresponding activities have been measured under high inducing conditions (0.5% galactose added to a 2% raffinose background) by flow cytometry (green bars) or β-Galactosidase CPRG assay (black bars) depending on the expression range. Expression was normalized with respect to wild type P_{GAL1} expression (WT) and error bars show standard deviations from two independent experiments comprising each at least 3 replicates. Native binding site sequences are denoted by the name of the gene containing them in its promoter sequence and in case a promoter contains more Gal4p binding sites then they are numbered (see Appendix I for sequences). 17mer corresponds to a high affinity artificial version of the consensus sequence (according to [25]) and binding sites situated to its right in the bar plot are artificial operators found in the literature ([26, 27]). The last sequence on the right side of the bar plot corresponds to the native binding site of the $GAL6$ gene with a single nucleotide mutation in the first position ($C \to A$). The experiments were obtained with strains and plasmids detailed in Tables 4, 5.

4.2 Systematic Study on the Effect of Multiple Activator Binding Sites in Promoters
Results

4.2.4.3.2 Measurement: Multiple $Gal4pBS$ Library

With the single operator library of $Gal4pBS$ inserted in the fourth position of the $GAL1$ promoter, single operator contributions could be measured and further compared to the expression of constructs constituting multiple operator library. In order to measure their expression deviation when present in multiple copies in P_{GAL1}, each of the $Gal4pBS$ selected during the calibration phase has been used to replace all of the four Gal4p binding sites present in the native P_{GAL1} configuration (where the single operator library corresponds to the configuration indicated as $[Gal4pBS]1$ and the multiple operator library to $[Gal4pBS]4$ in the schemes represented in Fig.39). For both calibration and measurement phase of the natural P_{GAL1} system the expression range of all the components constituting single and multiple operator libraries was expressed as a function of the wild type P_{GAL1} expression level (configuration indicated as $[WT]$ in the scheme of Fig.39). As previously, multiple *versus* single operators have been correlated through parametric representation together with the corresponding reference system and the outcome shown in Fig.48.

The contribution of multiple activator binding sites in P_{GAL1} activation was predominantly consisting of a more-than-additive increase in absolute expression level with respect to single binding sites. Consistent with the obtained saturation curve, the high induction conditions applied to the $GAL1$ promoter library were reflected by wild type P_{GAL1} activation exposed to the same conditions (blue star in Fig.A3).

4.2 Systematic Study on the Effect of Multiple Activator Binding Sites in Promoters
Results

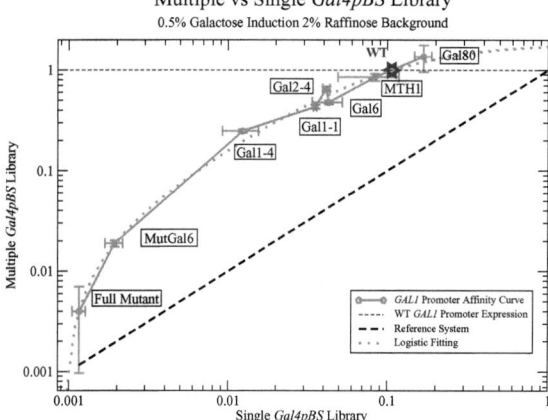

Figure 48: Measurement phase of multiple operator library. The x-axis represents the range of expression covered by the single operator library while on the y-axis by the multiple operator library. Measurement curves of the multiple operator library are obtained by combining expression values of x and y-coordinates under GAL genes inducing conditions (0.5% galactose added to a 2% raffinose background) while reference systems are constructed by the same procedure but comparing single *versus* single operator library, thus giving a diagonal line (black dashed line). Blue dashed line indicates the range of activity of the wild type promoter under same induction conditions as the promoter libraries. Green dotted line represents a fitted function to the dataset (standard logistic function (Eq.59) where the resulting parameters are: $A0 = -0.37$; $A1 = -0.79$; $A2 = -2.27$; $A3 = 1.76$). The experiments were obtained with strains and plasmids detailed in Tables 4, 5, 6.

4.2 Systematic Study on the Effect of Multiple Activator Binding Sites in Promoters
Results

4.2.4.4 Discussion

The approach presented in this section has been first validated by making use of a synthetic test system consisting of rtTA–*tet*-operators in order to assess the reliability of the technique under experimental conditions allowing external control. The technique has been thereby explored under different aspects and theoretical concepts could be verified experimentally. Subsequently, the technique has been applied to Gal4p-*Gal4pBS* interaction in the *GAL1* promoter, a system of relevance for gene regulation in which control mediated by external inducers cannot be efficiently used to study promoter activation. So, *GAL1* promoter activation profile could be accessed under minimally invasive experimental conditions.

Using the present approach, it has been possible to identify activation response of the two studied systems over a broad range of induction and the resulting curves were expressed parametrically to highlight deviations occurring by the presence of multiple with respect to single activator binding site sequences. With this method, activation curves describing promoters containing multiple activator binding sites are presented in terms of activation gain with respect to single binding site contributions. In order to compare the two analyzed systems, their corresponding dynamical ranges of activation with respect to single operator systems have been represented in the same figure (see Fig.49, left panel). In addition, parametric curves resulting from the natural system approach were fitted to a standard four-parameter logistic equation (Eq.59) (see dotted lines in Fig.45, Fig.48 and left panel of Fig.49) and the resulting optimally fitting functions used to compute logarithmic sensitivities represented graphically as a function of single operator activation (see Fig.49, right panel).

The contribution of multiple activator binding sites in P_{GAL1} activation was predominantly consisting of an increase of absolute expression level. With respect to single operator constructs, the effect of multiple operators in *GAL1* promoter resulted in rising system's response more homogeneously compared to the *tet*-operator test system. The *tet*-operator system displayed indeed a much sharper activation profile and was confined to a limited excursion range with narrower sensitivity peak. The results showed that the *GAL1* promoter is very reactive to low induction levels, but monotonously increasing its response in a proportionate manner at higher activity. P_{GAL1} activity ranges over a wide degree of induction maintaining a relatively constant response pattern at roughly 10-fold higher level with respect to the single operator system. The *GAL1* promoter is thereby able to integrate inputs spanning wide induction levels and convert them into output covering broad response range. In support to the present data interpretation, applying same inducing conditions to wild type P_{GAL1} indicates high degree of promoter activation. Since the experimental conditions to which the *GAL1* promoter library system is exposed to correspond to high induction (2% raffinose and 0.5% galactose), wild type P_{GAL1} activation is consistent to the boundaries of the natural system expression (Fig.49 left panel, blue star).

Several aspects revealed by the natural system approach can be connected to the experiments carried out in previous sections. Common points relating the systems studied with different approaches are discusssed in the following paragraphs in attempt to make the picture of the outcome of this first part of the work more interconnected. A common point relating present findings to previous studies of this part consists of the observation of quick system's reaction to changes in induction conditions combined with response

4.2 Systematic Study on the Effect of Multiple Activator Binding Sites in Promoters
Results

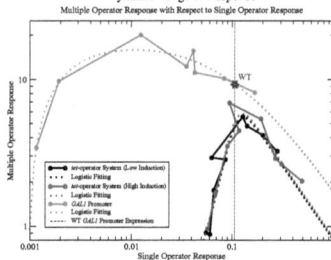
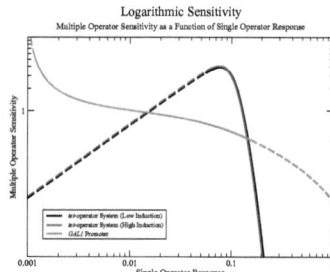

Figure 49: Activation profile of *tet*-operator and *GAL1* promoter systems. The effect of multiple operators (four in both cases) is represented with respect to corresponding single operator response. Left panel: The dynamic range of activation with respect to single operator systems has been computed by dividing datapoint values and fitted functions (from Fig.45 and Fig.48) by the reference system. The blue star indicates the level of expression corresponding to the wild type *GAL1* promoter exposed to same inducing conditions as the *GAL1* promoter library (2% raffinose and 0.5% galactose). Right panel: The curves fitted to parametric datasets have been used to compute logarithmic sensitivities of the two systems as a function of single operator response. Continuous lines indicate experimentally accessible range, dotted lines are extrapolated based on the fitted function.

4.2 Systematic Study on the Effect of Multiple Activator Binding Sites in Promoters
Results

scalability related to the presence of multiple activator binding sites (see section 'Semi-Synthetic Systems', Fig.40 and Fig.41). Similarly to the outcome of the natural system, using the semi-synthetic approach, *GAL1* promoter containing one or multiple activator binding sites rapidly reacts to early concentration changes. However, in case of the fully substituted P_{GAL1}, the presence of four operators rises its expression to a level roughly 10-fold higher than the corresponding single operator. Moreover, investigations of *GAL1* promoter derivatives under natural and semi-synthetic experimental setup indicate high reactivity to external induction even when driven by single operators. When gradually activated by a synthetic, externally tunable module, single operator P_{GAL1} reactivity was comparable to multiple operator one (roughly same K_D and n_H values, see Fig.40) while considerable difference was observed in the expression level scaling according to the number of activator binding sites. Similarly, when constant high induction was applied to the P_{GAL1} single operator library of variable affinity, high promoter activation was measured indicating that even when affinity for Gal4p was decreased by nucleotide mutations, single operator constructs could induce strong responses (see Fig.47). This effect is therefore unlikely to be due to binding cooperativity and operator affinity but rather to rely on activator interactions with the transcriptional machinery. The addition of multiple binding sites resulted in effect with both experimental approaches in a homogeneous increase of the activation level without changing qualitatively its activation profile compared to the activation curve resulting from the single activator site. At low induction, the multiple operator natural system quickly reacts in a switch-like manner by setting its activity to high levels, further increase in operator affinity simply scales proportionally the resulting promoter response in a graded way. The expression level reached by the multiple operator natural system is around 10-fold the corresponding single operator response and is consistent with measures performed previously when applying saturating induction conditions to *GAL1* promoters containing different number of operators (see Fig.41).

We propose that the role of multiple activator binding sites in the promoter of *GAL1* and possibly in the coregulated genes is mainly responsible for modulating absolute expression values rather than encoding intrinsic mechanisms of binding cooperativity. Scaling promoter response allows to quantitatively regulate transcription of different genes maintaining their activity synchronized. This effect displays more-than-additive features with respect to single operator systems characteristic of synergic activation and occurs at the level of the transcriptional machinery organization. The present situation exposes common points existing between results obtained by studying *GAL1* promoter regulation with semi-synthetic and natural approaches, and partially contrasting aspects concerning the full-synthetic approach. While the first two methodologies indicate strong promoter activation quickly reacting to low induction variations, full-synthetic approach applied to *GAL1* promoter displays an activation pattern that only partially concords with them (see section 'Full-Synthetic Systems'). Activation of promoters containing two to four activator binding sites in the full-synthetic setup is in accord with the mode of activation observed in the two complementary studies, however, the activation profile of P_{GAL1} containing a single *tet*-operator exhibits substantial differences. The scalability effect observed upon addition of activator binding sites in *GAL1* promoter is indeed present also in the full-synthetic system concerning constructs incorporating more than a single operator (see Fig.24), however, single *tet*-operator P_{GAL1} is less reactive to changing induction presenting a significantly different saturation curve.

4.2 Systematic Study on the Effect of Multiple Activator Binding Sites in Promoters
Results

A possible explanation for the observations resulting from the systematic analysis carried out with help of the three applied techniques, full-synthetic, semi-synthetic and natural approach, could have its basis on gene specific effects that confer the ability of Gal4p to promote efficient activation even if present individually. This capacity could perhaps rely on specific recruitment of cellular components with the consequence of converting Gal4p target promoter into a more receptive state of higher inducibility. Additional activator binding sites could thereby improve transcriptional efficiency by facilitated transcriptional initiation complex formation and result in higher gene expression level. Since concerning rtTA–tet-operators system the presence of a second binding sites had the capacity of relieving promoters from states of lower inducibility, as discussed in section 'Full-Synthetic Systems' regarding the $P_{tetOinEGT2}$ and $P_{tetOinCYC1}$ systems, the same mechanism could potentially be responsible for $P_{tetOinGAL1}$ activation profile conversion. A conserved feature consistent among all presented experimental setups is that, differently from the two other systems, $P_{tetOinGAL1}$ and wild type P_{GAL1} support the scaling effect that homogeneously increases its expression observed in the presence of multiple activator binding sites. In contrast, the fact that $GAL1$ promoter activation determined by single activators differs between full-synthetic system on one side, and semi-synthetic and natural systems on the other side, is reminiscent of experiments carried out in section 'Context-Dependent Effects on Gene Regulation' where the role of native elements taking part to the process of gene regulation through gene specific effects has been studied. Synthetic activators consisting of externally tunable rtetR DNA binding domain were fused with transcriptional activation domains of native activators or, alternatively, native activators were allowed to bind the target promoter under study in order to integrate gene specific interactions potentially occurring with the transcriptional machinery in an experimental framework consisting of externally inducible activator binding. The resulting effect of native elements allowed to participate to the synthetically driven process of promoter activation controlled externally highlighted their role in altering promoter properties. Firstly, similarly to differences observed in this section between synthetic, semi-synthetic and natural systems, promoter reactivity could be modulated based on the type of activator used to mediate induction, which in turn resulted to integrate same inputs differently. Under same induction conditions, fusion proteins containing different transcriptional activation domains could thereby activate same promoter sequences with different activation profiles. Secondly, change in promoter response could be observed upon influence of gene specific elements allowed take part of the process according to their native regulation. The state of the same promoter regulated by multiple activator binding sites could in effect be altered by allowing single native activator, and the consequent interplay with its specific components, to interact with it independently on the externally inducible activator binding.

In summary, combining the different approaches we could reveal that promoter saturation functions of a certain graded induction stimulus could be altered in their activation profile based on context dependent factors and consequently display different behavior towards same stimuli. This effect could also be achieved independently on the DNA sequence thus highlighting the role of promoter reorganization induced by native components of the cellular machinery that affect gene regulation *in vivo*. Some of the aspects of promoter regulation could be captured and understood with help of the relatively simplistic approach employed in full-synthetic systems (such as the scalability effect of expression values observed for $GAL1$ promoter regulation by multiple

4.2 Systematic Study on the Effect of Multiple Activator Binding Sites in Promoters
Results

binding sites), while other aspects required more laborious technique depicted by the natural approach in order to be accessed in a more detailed way (like single operator based reactivity of *GAL1* promoter). While regulation of synthetic systems could be equally successfully studied using both full-synthetic or natural approach, regulation of endogenous promoters could only be precisely determined with the latter technique that allowed to take into account the involved native specific effects. Indeed, due to the isolation degree of full-synthetic systems with respect to cellular components, both approaches demonstrated to correlate very precisely according to theoretical predictions. In contrast, the role of native components taking part to the regulatory process and commonly completely neglected in studies of *in vitro* protein-DNA interactions resulted to play determining role in shaping endogenous promoter response.

Promoters studied in the present work could display substantially different activation patterns depending on molecular partners recruited additionally to the basal ones. These cell's endogenous factors have the capacity to support particular features pointing out the higher degree of complexity underlying the process of gene regulation and provides evidence for the fact that both promoter sequences *and* activators share responsability in the way a stimulus is integrated in promoter regulatory sequences to be further interpreted and converted into a genetic response. Activation curves resulting from our study are therefore composite functions combining higher level properties of promoter regulation with respect to simpler *in vitro* studies.

5 Conclusions and Future Work

5.1 Conclusions

It is commonly assumed that gene regulation occurs in an all-or-none fashion. However, graded gene expression is necessary to display proportionate response to stimuli and represents a fundamental aspect to appropriately adapt to a changing environment. Indeed, it's obvious that for certain processes a binary response is inadequate and more fine tuned regulation is necessary to avoid production excess or insufficiency of the required components participating to the response mechanism. It is known that certain transcriptional regulators can bind different regulatory sequences in the genome in order to modulate activity of several genes. Pleiotropy is a typical regulatory feature present among different organisms and combined with the fact that different target genes for a common regulator can contain variable number of binding sites in their regulatory sequence hints for a potentially relevant role played by synergic effects between regulators. In enzyme kinetics and ligand binding, multiple ligand binding to receptors is commonly associated with binding cooperativity and thus with switch-like response patterns while single ligand binding with graded responses. The current work has been focused on the role of multiple binding sites present in promoter sequences with respect to single binding and the consequences on systems' sensitivity and reactivity. If in biological systems a common mechanism to achieve switch-like response induced by gradually increasing stimulus relies on positive feedback loops, binding cooperativity provides additional means for modulating gene activation profiles without the requirement of regulator's positive retroaction. Gene regulatory sequences could integrate and interpret same signals in different modes by simply embedding in the promoter architecture a signal conversion mechanism supporting from single to multiple regulator binding and thereby modulating their activation profile with different degrees of sensitivity from graded, dose-dependent induction to more switch-like to binary response.

In order to understand function and properties of gene systems regulated by promoters containing single *versus* multiple activator binding sites and to shed light on their corresponding gene activation profile, we made use of modern techniques of synthetic biology and genetic engineering to be able to access parameters describing the process of gene regulation in living cells. Inspired by the experimental evidence hinting for a relevant role played by the context specificity involved in the regulation of cell cycle genes containing multiple activator binding sites in their promoters, the effect of methodically varying the number of activator binding sites has been initially studied using synthetic activator-binding site partners (rtTA–*tet*-operators) based on bacterial elements inserted in yeast promoters and induced with doxycycline. This system (full-synthetic system, according to the previously used denomination) revealed important aspects of promoter sequences especially in the way they integrate multiple activator based inputs. It resulted that inputs mediated by same the activator-operator pair was converted differently into gene activation responses based on the surrounding promoter context. By progressively increasing the number of rtTA binding sites (*tet*-operators), different promoters could display significantly diverse gene activation profiles outlining the role of core promoter sequences to interpret induction stimuli. By interpreting experimental results with mathematical tools commonly used to study enzymatic and biomolecular reaction systems it resulted that the difference both on the qualitative and quantitative level was largest when comparing gene activation based on lower binding events. Actually, by increasing

5.1 Conclusions

the number of activator binding sites from one to two/three (depending on the cases) a very large impact on gene activation was observed compared to subsequent additional binding sites.

An extreme case highlighting the role of low order binding is represented by *GAL1* core promoter: Comparing the activation profiles obtained by gradually inducing *GAL1* promoters containing variable *tet*-operators revealed particular features of dose-response curves differing substantially between promoters regulated by single *and* promoters regulated by multiple operators. *GAL1* promoter activation could in fact exhibit two different types of activation profiles: A low sensitivity profile presenting graded response when regulated by a single activator binding site and a high sensitive response when regulated by two or more activator binding sites. Moreover, upon further increase of activator binding sites the same activation pattern is conserved and saturation curves with same properties as for two binding sites are observed with the only difference consisting of being homogeneously scaled to higher gene expression levels. Because of this very pronounced difference in the mode of activation and together with the fact that, in addition to *GAL1*, several other genes that participate to the galactose metabolism contain different number of binding sites for the common native activator Gal4p, we extended the study using an experimental setup that allowed to externally vary the Gal4p expression and thereby to obtain saturation curves of *GAL1* promoter regulated by its native activator instead of the synthetic activator rtTA. According to this methodology (semi-synthetic system, based on the previously used denomination), a synthetic, externally tunable module has been used to replace the native *GAL4* gene allowing to adjust the activator production and to observe the corresponding target gene activation. Thus, different versions of the same target promoter (*GAL1* promoter) have been exposed to variable Gal4p concentrations by focusing on the role of number and affinity of Gal4p binding sites. The obtained saturation curves were characterized by switch-like responses displayed by all of the studied versions of *GAL1* promoter that on one hand, consistently to the results obtained previously, were scaled homogeneously accordingly to the number and affinity of activator binding sites, but on the other hand presented highly sensitive activation even when regulated by single activators. This indicated that the full-synthetic experimental setup could appropriately represent the natural process for some aspects however lacking the capacity to integrate some features potentially involving recruitment of gene specific elements that influence activation based on single activators. Since the semi-synthetic method was supposed to be significantly invasive because potentially perturbing external branches of the cellular metabolism as a consequence of the interference with the expression of native regulators, we turned to a method consisting of synthetic promoter libraries of variable affinity. This method (natural system, according to the previously used denomination) didn't imply modifications in cellular regulation and allowed to perform experiments in wild type cells in order to study the process of gene regulation under native conditions. It resulted that employing the natural approach allowed to obtain saturation curve describing *GAL1* promoter over a broad range of activation and suggested that the promoter is generally very reactive to low induction levels, but monotonously increasing its response in a proportionate manner with higher degree of induction. Hence, *GAL1* promoter displayed switch-like response at low activity but graded response when exposed to higher induction levels.

In addition to the resulting points described above, another experimental evidence supports the idea arguing in favor of the capacity of promoters regulated by single oper-

5.1 Conclusions

ators to display sensitive response. In fact, the vast majority of the naturally occurring binding site sequences recognized by Gal4p in native conditions have strong activation potential even with low affinity. Moreover, also in this case, the presence of multiple activator binding sites mainly contributed to scale the response level in a more than additive manner with respect to single operator construct indicating synergy in the recruitment of gene specific components interacting with bound activators. This effect seems thereby not be related to binding site affinity and binding cooperativity but rather to cooperativity at the level of activator transcriptional machinery interaction and supports the idea that eukaryotic activators can cooperate not by directly interacting but by simultaneously touching some components of the transcriptional machinery [28, 29]. In addition to the role of low order binding events, native activators allowed to interact with synthetically tunable promoters also contributed to change the responsiveness and the capacity of promoters to integrate stimuli (see section 'Context-Dependent Effects on Gene Regulation'), supporting the idea that promoters remain in a state of lower activity and can be converted into a state of higher reactivity by recruitment of gene specific elements. Alternatively, as suggested by full-synthetic systems, in absence of gene specific effects a certain number of bound activators is required to unblock its state. Mechanisms such as increased polymerase trafficking or destabilization of structured DNA upon protein-DNA interaction could explain the obtained results (see 'Part II'). Binding cooperativity as defined in enzyme kinetics supported by protein-protein interaction between adjacently bound activators has been ruled out for the full-synthetic system and is also apparently not driving *GAL1* promoter's reactive response.

The technique developed for this work shows that selected genes can be efficiently studied using promoter libraries providing means to obtain dose-response curves over a broad range of activation under minimally invasive experimental conditions. Knowing the consensus sequence recognized by an activator allows one to construct a series of graded alterations in order to assess the effect of single and multiple activators in integrating induction stimuli and convert them into gene expression response. The technique proposed in this work allows to access such information assuming precise knowledge of the consensus sequence for a given activator, however modern techniques of *in vitro* and *in vivo* directed evolution offer the possibility to create and optimize promoter libraries of variable affinity for any genes providing the basis for the methodology presented in the current work [20, 19, 21, 22]. The applicability of the technique is confirmed for the studied system from *E.coli* to yeast and is potentially extendable to higher eukaryotes. A possible bottleneck in the technique is represented by weakly expressing gene systems and consequent insufficient signal intensity during detection analysis. The method can in effect require detection systems of higher sensitivity with respect to fluorescent reporter gene expression (LacZ assay, real-time PCR).

Protein-DNA binding is classically studied with techniques of *in vitro* experimentation and mathematical modeling is used to interpret experimental data. Saturation functions retrieved in this way often fail to represent the actual process occurring *in vivo* due to the absence of cellular components participating to the process under natural conditions. Saturation functions are alternatively evaluated *in vivo* by sampling a continuum of gene expression through few discrete datapoints anyway providing an incomplete picture of the gene activation process [19]. The proposed technique provides instead continuous access of gene expression as a function of graded induction and allows to take into consideration genetic and cellular aspects often relevant for the description of the studied system. Its application could range from the characterization of mechanisms underlying activation of selected genes to synthetic experimental setup in which precise tuning of gene expression is often required to obtain desired properties in genetic networks. Elements of such promoter libraries could thereby be employed to improve pathway behavior and consequently optimize the design of artificial networks.

5.2 Future Work

An interesting open question consists in applying the same procedure according to the natural system design employed to characterize *GAL1* promoter to other genes coregulated by Gal4p and to determine if the role of multiple binding sites present in P_{GAL1} implies comparable consequences in the regulation of other promoters involved in the galactose network. This could help to understand why metabolic GAL genes are mainly regulated through multiple binding sites whereas genes encoding for regulators by single binding sites. Candidates for such work are genes regulated by multiple Gal4p binding sites on one hand (*GAL2*, *GAL7*) and *GAL3* promoter on the other. *GAL3* and *GAL1* are in fact paralogues arising from a single bifunctional ancestral gene still present in the yeast *Kluyveromyces lactis* and encoding for a bifunctional protein combining the galactokinase activity of *Saccharomyces cerevisiae*'s Gal1p and the galactose sensing of Gal3p. *GAL3* and *GAL1* genes of modern *Saccharomyces cerevisiae* have thereby evolved from an ancestral sequence known to have been subject to a process of gene duplication and subsequent subfunctionalization. The corresponding promoter have common origin and contain one and four Gal4p binding sites, thus representing ideal candidates to be compared over their range of activity. Moreover, the single bifunctional ancestral gene is still present in *Kluyveromyces lactis* and can be studied with the same method in order to compare its activation profile with *GAL3* and *GAL1* genes form *Saccharomyces cerevisiae* [7]. In this evolutionary scenario it has been shown that implications of activator binding through in/out helical phasing play an important role in optimizing gene regulation [7, 109]. However those studies have been carried out by evaluating gene expression through few discrete datapoints characterizing extremes of the real saturation function describing the gene system. Using the natural approach proposed in the present work would instead allow to compare the behavior of such promoters over a broad range of activity.

Another open question remains about the positioning role of the binding sites for the *Gal4pBS* in *GAL1* promoter. The difference in using operator substitutions in the proximal position or in the distal and the relative combinations has not been methodically studied but experimental evidence suggests that some mechanism acting at that level can play a role in the mode of *GAL1* promoter activation (see Appendix I). Furthermore, in order to precisely consider the more important role of activator binding when occurring at low binding order, a common point relating the systems studied in this part, promoters with intermediate number of binding sites can be further studied using the natural system approach and thus the role of low order binding events better characterized.

6 Appendix I

6.1 Promoter Sequences

6.1.1 $P_{tetO2inSIC1}$

GATACGCGTTCCCTATCAGTGATAGAGACTGAAAATTTCCCTATCAGTGA
TAGAGATTGTTCGCGTCGCAACTTTTGATATTTATAAAAAGAGAGAAGGC
ATATCCAAAGGATAAGGTATTGTCTTTGTAAATCAATCAACCAAACCTCT
ACGGAATTTTGACCCTTGAAGCAGGGACTATTACACGAAA

$tetO2$ type tet-operators are underlined.

6.1.2 $P_{tetO5inEGT2}$

ACTAATATTCCCTATCAGTGATAGAGAAGTAACCTCTTTAGTAACCTCTT
TAGTGAATCCCTATCAGTGATAGAGATTATTTTCCTAATTCGGACGCGCT
GGCTCCGCGCCGTGAGGAAAATCCCTATCAGTGATAGAGACTAATTTATC
CCTATCAGTGATAGAGACTTCCGAAAAGGTACTCAAGCTTTTATCCCTAT
CAGTGATAGAGATGCTAAGAATACGATTTAGTACTTCCTTTTTAATGTGG
CTTGTTTTTTTTTATTCGTCCATAACAGTACAAATATAAAAGCAGAATA
CATTCTTGCCGACCTAAAACTTGTGTCCTTAGACTCAAACTTTCTGATGG
TTTTCTTTCAGTAGAATCAACATAAACAAGCATAATATAAAACACTACA

$tetO2$ type tet-operators are underlined.
Swi5p/Ace2p binding site underlined twice.

6.1.3 $P_{tetO6inEGT2}$

ACTAATATTCCCTATCAGTGATAGAGAAGTAACCTCTTTAGTAACCTCTT
TAGTGAATCCCTATCAGTGATAGAGATTATTTTCCTAATTCGACTCTATC

<u>ATTGATAGAGT</u>CGCCGTGAGGAAAA<u>TCCCTATCAGTGATAGAGA</u>CTAATT
TA<u>TCCCTATCAGTGATAGAGA</u>CTTCCGAAAAGGTACTCAAGCTTTTA<u>TCC
CTATCAGTGATAGAGA</u>TGCTAAGAATACGATTTAGTACTTCCTTTTTAAT
GTGGCTTGTTTTTTTTTTATTCGTCCATAACAGTACAAATATAAAAGCAG
AATACATTCTTGCCGACCTAAAACTTGTGTCCTTAGACTCAAACTTTCTG
ATGGTTTTCTTTCAGTAGAATCAACATAAACAAGCATAATATAAAACACT
ACA

tetO2 type *tet*-operators are underlined.
tetO1 type *tet*-operators are underlined twice.

6.1.4 $P_{tetO1inCYC1}$

TAATTCGCGCCACTTCTAAATAAGCGAATTTCTTATGATTTATGATTTTT
ATTATTAAATAAGTTATAAAAAAAATAAGTGTATACAAATTTTAAAGTGA
CTCTTAGGTTTTAAAACGAAAATTCTTGTTCTTGAGTAACTCTTTCCTGT
AGGTCAGGTTGCTTTCTCAGGTATAGCATGAGGTCGCTCTTATTGACCAC
ACCTCTACCGGCAGATCAATTCCTCGA<u>TCCCTATCAGTGATAGAGA</u>GTCG
ACAAAGTCGAGTTTCTCGATCGAGACCACTGCATGCATGTGCTCTGTATG
TATATAAAACTCTTGTTTTCTTCTTTTCTCTAAATATTCTTTCCTTATAC
ATTAGGTCCTTTGTAGCATAAATTACTATACTTCTATAGACACGCAAACA
CAAATACACACACTAAATTACCGGATCAATTCGGGGGATCC

tetO2 type *tet*-operators are underlined.

6.1.5 $P_{tetO2inCYC1}$

TAATTCGCGCCACTTCTAAATAAGCGAATTTCTTATGATTTATGATTTTT
ATTATTAAATAAGTTATAAAAAAAATAAGTGTATACAAATTTTAAAGTGA
CTCTTAGGTTTTAAAACGAAAATTCTTGTTCTTGAGTAACTCTTTCCTGT
AGGTCAGGTTGCTTTCTCAGGTATAGCATGAGGTCGCTCTTATTGACCAC
ACCTCTACCGGCAGATCAATTCCTCGA<u>TCCCTATCAGTGATAGAGA</u>GTCG
ACAAAGTCGAGTTTCTCGA<u>TCCCTATCAGTGATAGAGA</u>GTCGACAAAGTC
GAGTTTCTCGATCGAGACCACTGCATGCATGTGCTCTGTATGTATATAAA
ACTCTTGTTTTCTTCTTTTCTCTAAATATTCTTTCCTTATACATTAGGTC
CTTTGTAGCATAAATTACTATACTTCTATAGACACGCAAACACAAATACA
CACACTAAATTACCGGATCAATTCGGGGGATCC

Appendix I   Appendix I   Appendix I

tetO2 type *tet*-operators are underlined.

6.1.6 $P_{tetO7inCYC1}$

TAATTCGCGCCACTTCTAAATAAGCGAATTTCTTATGATTTATGATTTTT
ATTATTAAATAAGTTATAAAAAAAATAAGTGTATACAAATTTTAAAGTGA
CTCTTAGGTTTTAAAACGAAAATTCTTGTTCTTGAGTAACTCTTTCCTGT
AGGTCAGGTTGCTTTCTCAGGTATAGCATGAGGTCGCTCTTATTGACCAC
ACCTCTACCGGCAGATCAATTCCTCGATCGAGTTTACCAC<u>TCCCTATCAG</u>
<u>TGATAGAG</u>AAAGTGAAAGTCGAGTTTACCAC<u>TCCCTATCAGTGATAGAG</u>
AAAAGTGAAAGTCGAGTTTACCAC<u>TCCCTATCAGTGATAGAG</u>AAAGTGA
AAGTCGAGTTTACCAC<u>TCCCTATCAGTGATAGAG</u>AAAGTGAAAGTCGAG
TTTACCAC<u>TCCCTATCAGTGATAGAG</u>AAAGTGAAAGTCGAGTTTACCAC
<u>TCCCTATCAGTGATAGAG</u>AAAGTGAAAGTCGAGTTTACCAC<u>TCCCTATC</u>
<u>AGTGATAGAG</u>AAAGTGAAAGTCGAGCTCGGTACCCTATGGCATGCATGT
GCTCTGTATGTATATAAAACTCTTGTTTTCTTCTTTTCTCTAAATATTCT
TTCCTTATACATTAGGTCCTTTGTAGCATAAATTACTATACTTCTATAGA
CACGCAAACACAAATACACACACTAAATTACCGGATCAATTCGGGGGATC
C

tetO2 type *tet*-operators are underlined.

6.1.7 $P_{tetOinEGT2}$

GctagcGATactagtGATggcgcgccGATcctaggGATtctagaTTTGAT
GCTAAGAATACGATTTAGTACTTCCTTTTTAATGTGGCTTGTTTTTTTTT
TATTCGTCCATAACAGTACAAATATAAAAGCAGAATACATTCTTGCCGAC
CTAAAACTTGTGTCCTTAGACTCAAACTTTCTGATGGTTTTCTTTCAGTA
GAATCAACATAAACAAGCATAATATAAAACACTACA

Lower case sequences are the restriction sites *NheI, SpeI, SgsI, AvrII, XbaI* used to build promoter libraries and different constructs containing from one to five *tetO1* type *tet*-operators.

125

6.1.8 *GAL1* Promoter

ATATTGAAGTA<u>CGGATTAGAAGCCGCCGAGCGGGCGACAGCCCTCCGACG</u>
<u>GAAGACTCTCCTCCG</u>TGCGTCCTCGTCTTCACCGGTCGCGTTCCTGAAAC
GCAGATGTGCCT<u>CGCGCCGCACTGCTCCG</u>AACAATAAAGATTCTACAATA
CTAGCTTTTATGGTTATGAAGAGGAAAAATTGGCAGTAACCTGGCCCCAC
AAACCTTCAAATTAACGAATCAAATTAACAACCATAGGATGATAATGCGA
TTAGTTTTTTAGCCTTATTTCTGGGGTAATTAATCAGCGAAGCGATGATT
TTTGATCTATTAACAGATATATAAATGGAAAAGCTGCATAACCACTTTAA
CTAATACTTTCAACATTTTCAGTTTGTATTACTTCTTATTCAAATGTCAT
AAAAGTATCAACAAAAAATTGTTAATATACCTCTATACTTTAACGTCAAG
GAGAAAAAACTATA

The four different native Gal4p binding sites present in the wild type *GAL1* promoter are underlined. These sequences are replaced by *tetO2* type *tet*-operators according to Fig.20 or alternatively substituted with other artificial Gal4p binding sites or belonging to other target promoters among the GAL genes (see Fig.39). Mutant Gal4p binding sites have been obtained by exchanging purine with pyrimidine in the consensus sequence for the Gal4p binding sites ($CGG[N]_{11}CCG \rightarrow GCC[N]_{11}GGC$).

6.2 tet-Operators

Two genes, *tetA* and *tetR*, are widely present among gram-negative bacteria and confer tetracycline resistance (which is the most common form of antibiotic resistance) by promoting active drug efflux. The two genes are divergently oriented and a central overlapping regulatory region drives their expression. *tetA* codes for a tetracycline/metal-proton membrane antiporter (TetA) and *tetR* codes for a gene regulator (TetR) which is a dimeric DNA binding protein and is responsible for the repression of *tetA* and *tetR* genes in absence of tetracycline. Moreover, the antibiotic activity of tetracycline acts at the protein production level inhibiting translation and therefore the mechanism of resistance is supposed to be regulated in a sensitive way. Occupation of *tetO1* operator leads to repression of both genes while occupation of *tetO2* operator causes repression of *tetA* but doesn't interfere with expression of *tetR* (see Fig.50) [112, 113].

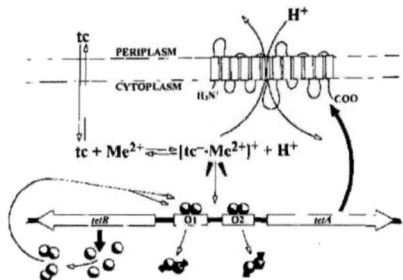

Figure 50: Scheme representing the genetic mechanism of regulation responsible for tetracycline resistance in gram-negative bacteria (taken from [112]).

The sequences of *tetO1* and *tetO2* operator are ACTCTATCATTGATAGAGT and TCCCTATCAGTGATAGAGA, respectively. The *tet*-operator library has been constructed according to information found in [23] and Table 1 shows the different point mutations (represented by capital letters) applied to one half of the *tetO1* operator palindromic sequence thereby causing a decrease in binding affinity for the tetR DNA binding domain.

Operator	Activity [23]	Sequence
Wild Type	100%	actctatcattgatagagt
9C	68%	actctatcattgatagagC
8C	41%	actctatcattgatagaCt
5T	22%	actctatcattgatTgagt
6T	13%	actctatcattgataTagt
6C	10%	actctatcattgataCagt
2T	6%	actctatcattTatagagt
3T	2%	actctatcattgTtagagt
2C	0%	actctatcattCatagagt

Table 1: *tet*-operator single nucleotide mutations constituting *tet*-operator libraries. The decrease in affinity is reported as fractional activity with respect to the wild type sequence.

Description	Plasmid Name	Yeast Strain	Reporter
Wild Type	pAnt36	YAntH191.5,6,7,9,10,11	GFP
9C	pAnt37	YAntH192.1,3,5,10,11,12	GFP
8C	pAnt38	YAntH193.1,4,6,7,10,12	GFP
5T	pAnt39	YAntH194.1,2,4,5,10,12	GFP
6C	pAnt42	YAntH195.2,3,9,10	GFP
2T	pAnt43	YAntH196.1,3,4,6,8,10,11	GFP
2C	pAnt45	YAntH197.1,3,4,5,6,7,8,9	GFP
6T	pAnt132	YAntH209.2,4,5,8,9,10,12	GFP
3T	pAnt123	YAntH203.3,8,9,11,12	GFP

Table 2: Single *tet*-operator library.

Description	Plasmid Name	Yeast Strain	Reporter
Wild Type	pAnt56	YAntH216.1,2,3,6	GFP
9C	pAnt57	YAntH217.1,2,3,4	GFP
8C	pAnt58	YAntH218.1,2,6,7	GFP
5T	pAnt59	YAntH219.2,3,5,6	GFP
6C	pAnt60	YAntH220.2,3,5,10	GFP
2T	pAnt61	YAntH221.1,2,4,10	GFP
2C	pAnt62	YAntH222.1,2,4,5	GFP
6T	pAnt141	YAntH234.2,4,5	GFP
3T	pAnt143	YAntH235.2,3,5	GFP

Table 3: Multiple *tet*-operator library (4 binding sites).

6.3 Gal4p Operators

Native binding site sequences are denoted by the name of the gene containing it in its promoter sequence and in case a promoter contains more Gal4p binding sites then they are numbered. Concerning the denomination used in section 'Semi-Synthetic Systems', where *Gal4pBS* denotes general Gal4p binding site and the following binding sites have been used in that context: *GAL80BS*, *GAL1BS1*, *GAL1BS2*, *GAL1BS3*, *GAL1BS4* corresponding to the denomination used here and in section 'Natural Systems' as Gal80, Gal1-1, Gal1-2, Gal1-3, Gal1-4.

Operator	Sequence
Gal1-1	CGGATTAGAAGCCGCCG
Gal1-2	CGGGCGACAGCCCTCCG
Gal1-3	CGGAAGACTCTCCTCCG
Gal1-4	CGCGCCGCACTGCTCCG
Gal2-1	CGGCGGTCTTTCGTCCG
Gal2-2	CGGGGCGGATCACTCCG
Gal2-3	CGGAAAGCTTCCTTCCG
Gal2-4	CGGAGATATCTGCGCCG
Gal2-5	CGGATCACTCCGAACCG
Gal3	CGGTCCACTGTGTGCCG
Gal6	CGGCTGACAAGTCGCCG
Gal7-1	CGGACAACTGTTGACCG
Gal7-2	CGGAGCACTGTTGAGCG
Gal80	CGGCGCACTCTCGCCCG
GCY	CGGGGCAGACTATTCCG
MTH1	CGGGGAAATGGAGTCCG
MEL1	CGGCCATATGTCTTCCG
17mer	CGGAAGACTCTCCTCCG
Art.1	CGGGTGACCGCCCTCCG
Art.2	AGGAAGACTCTCCTCGG
Art.3	CGGAGGACTGTCCTCCG
Art.4	CGGAGGACGGTCCTCCG
wGal1-2	CGGGTGACAGCCCTCCG
wGal1-3	AGGAAGACTCTCCTCCG
mutGal6	AGGCTGACAAGTCGCCG

Table 4: *Gal4pBS*: DNA binding sites that bind Gal4p.

Description	Plasmid Name	Yeast Strain	Reporter
Gal1-1	pAnt100	YAntH129.6,1,2,3	GFP
Gal1-2	pAnt113	YAntH141.1,2,3	GFP
Gal1-3	pAnt114	YAntH142.1,2,3	GFP
Gal1-4	pAnt86	YAntH134.2,3,6,7	GFP
Gal2-1	pAnt95	YAntH124.2,7,8,9	GFP
Gal2-2	pAnt103	YAntH132.6,2,4,5	GFP
Gal2-3	pAnt104	YAntH133.7,5,6,13	GFP
Gal2-4	pAnt106	YAntH135.4,1,2,3	GFP
Gal2-5	pAnt107	YAntH136.6,1,2,4	GFP
Gal3	pAnt96	YAntH125.1,2,3,5	GFP
Gal6	pAnt110	YAntH139.9,1,5,6	GFP
Gal7-1	pAnt94	YAntH123.1,7,9,11	GFP
Gal7-2	pAnt102	YAntH131.4,3,5,10	GFP
Gal80	pAnt108	YAntH137.5,7,9,11	GFP
GCY	pAnt109	YAntH138.4,6,10,13	GFP
MTH1	pAnt116	YAntH151.1,2,3	GFP
MEL1	pAnt93	YAntH122.3,6,9,13	GFP
17mer	pAnt111	YAntH140.1,2,3	GFP
Art.1	pAnt98	YAntH127.1,4,9,10	GFP
Art.2	pAnt99	YAntH128.6,1,2,3	GFP
Art.3	pAnt91	YAntH120.2,4,5,6	GFP
Art.4	pAnt92	YAntH121.2,3,4,5	GFP
wGal1-2	pAnt97	YAntH126.2,3,5,6	GFP
wGal1-3	pAnt101	YAntH130.5,9,10,15	GFP
Gal1-1	pAnt144	YAntH226.1,2,6	LacZ
Gal1-4	pAnt140	YAntH225.1,2,3	LacZ
Gal2-3	pAnt145	YAntH227.1,2,3,4,5,6	LacZ
Gal2-4	pAnt146	YAntH228.3,4,5	LacZ
Gal6	pAnt151	YAntH233.1,5,6	LacZ
Gal7-1	pAnt150	YAntH232.1,3,5	LacZ
17mer	pAnt149	YAntH231.1,2,3	LacZ
Art.2	pAnt147	YAntH229.1,2,3,4,5,6	LacZ
mutGal6	pAnt173	YAntH281.4,5,6	LacZ
wGal1-3	pAnt148	YAntH230.1,2,3,4,5,6	LacZ

Table 5: Single operator (*Gal4pBS*) library.

Description	Plasmid Name	Yeast Strain	Reporter
Full Mutant	pAnt82	YAntH110.2,3,4	GFP
Wild Type	pAnt84	YAntH111.2,3	GFP
Gal1-1	pAnt154	YAntH254.4,10,11	GFP
Gal1-4	pAnt155	YAntH255.3,11,12	GFP
Gal80	pAnt156	YAntH256.6	GFP
Gal2-4	pAnt160	YAntH257.3,7,8	GFP
Gal6	pAnt161	YAntH258.4,5,10	GFP
MTH1	pAnt162	YAntH259.7,10,11	GFP
mutGal6	pAnt175	YAntH308.2,6,7	GFP

Table 6: Multiple operator ($Gal4pBS$) library (4 binding sites).

Description	Plasmid Name	Yeast Strain	Reporter
Gal1-1	pAnt188	YAntH317.2,3,7	GFP
Gal1-4	pAnt189	YAntH318.3,4,7	GFP
Gal2-4	pAnt187	YAntH316.2,3,6	GFP
Gal6	pAnt186	YAntH315.1,2,3	GFP
MTH1	pAnt185	YAntH314.3,7,8	GFP
mutGal6	pAnt190	YAntH319.7,9,11	GFP

Table 7: Double operator ($Gal4pBS$) library.

Description	Plasmid Name	Yeast Strain	Reporter
MGGM	pAnt119	YAntH297.4,6	GFP
GMMG	pAnt171	YAntH298.2,3,10	GFP
MWWM	pAnt120	YAntH299.1,4,9	GFP
GGGM	pAnt118	YAntH300,2.5.7	GFP

Table 8: $GAL1$ promoter with variable number and type of $Gal4pBS$. The denomination used in the description column is as follows: M (mutated $Gal4pBS$), G (Gal80 binding site), W (wild type binding sites). The first letter in the description corresponds to the most distal $Gal4pBS$, the last to the most proximal.

6.4 Comparison pAnt117 (pAnt27-derived (no *FIG1*-integration-locus)) and pAnt156 (pPR173-derived (with *FIG1*-integration-locus))

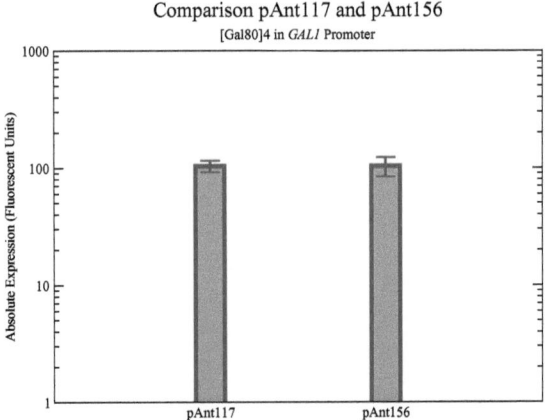

Figure 51: The promoter libraries for the study of *GAL1* promoter have been constructed using two different parent plasmids: pAnt117, from pAnt27 which is derived from a pRS306 plasmid. pAnt156 derived plasmids contain a *FIG1* integration locus in addition to the pRS306 elements. For all the experiments carried out in the present work genetic constructs were integrated in the yeast genome using the *URA* resistance gene, independently on the presence of the *FIG1* integration locus. The above plot shows the same library construct (*Gal80BS* substituting all four binding sites in the *GAL1* promoter) measured using the different parent plasmids.

6.5 Positional Effect of $Gal4pBS$ in $GAL1$ Promoter

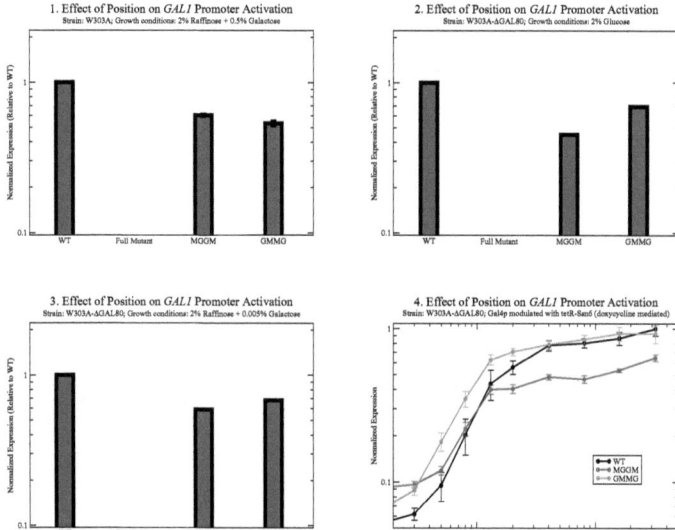

Figure 52: Position effect of $Gal4pBS$. All expression measures have been normalized with respect to the wild type (WT) expression, in the bottom figure right normalized to the expression value at maximal induction reached by the wild type. $GMMG$ and $MGGM$ indicate which operator of $GAL1$ promoter has been substituted with $GAL80BS$ corresponding to a letter G and operators that have been mutated denoted with a letter M. The denomination reflects the positioning of $Gal4pBS$ within the promoter configuration of P_{GAL1} in which the last letter corresponds to the proximal binding site. Under different condition the two substituted promoter versions display different activation levels. Top left: WT strain grown in 2% raffinose and 0.5% galactose. Top right: $\Delta Gal80$ strain grown in 2% glucose. Bottom left: $\Delta Gal80$ strain grown 2% raffinose and 0.005% galactose. Bottom right: $\Delta Gal80$, $SyntheticGAL4expression$ strain grown 2% raffinose and 0.5% galactose.

6.6 Feedback of *GAL1* Gene

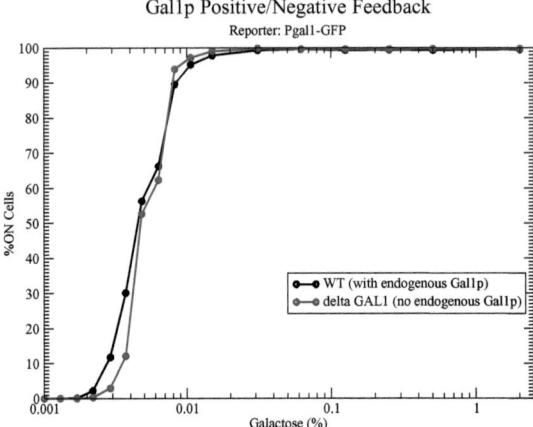

Figure 53: The feedback effect of the *GAL1* gene.

7 Appendix II

This part has been added separately in a late stage of the writing of the thesis. More experiments have been performed during the revision period of the thesis and the outcome is presented in this appendix. The main goal of this part is to complete the picture outlined in the main text concerning the role activator binding sites in the *GAL1* promoter. A series of additional constructs has thereby been made in order to verify the activation state of the *GAL1* promoter relying only on the presence of the two central Gal4p binding sites. Our previous investigations showed that the two central binding sites in the *GAL1* promoter were strong with respect to the two weak flanking ones. The same setup described in the section 'Natural Systems' regarding the study of the *GAL1* promoter consisting of four *Gal4pBS* has been transferred to the study of the *GAL1* promoter library activated only by the two central activator binding sites. The results shown in Fig.54 represent the library of four operator measured at the same time with the two operator library and other modified *GAL1* promoter constructs. The results show that a similar gene response is obtained between the two libraries in terms of sensitivity (see Fig.55). The main difference highlighting non-linearity in the gene activation profile can be observed in the scaling ratio between the curve traced by the two systems. The two systems react in similar fashion with the main difference being the absolute expression level: The double operator library improved expression 2/3-fold with respect to the single operator library while the multiple (four binding sites) improved gene expression roughly 15-fold.

The results presented here further support previous findings and shed light in a more detailed way on the role of multiple binding sites in the *GAL1* promoter. Cooperativity appears thereby to rely on interactions between activators and the transcriptional machinery rather than between activators. To further support this conclusions the expression of various *GAL1* promoter versions consisting of different number and types of *Gal4pBS* has been included in the same graphical representations showing their relative increase in absolute gene expression (blue symbols in Fig.54).

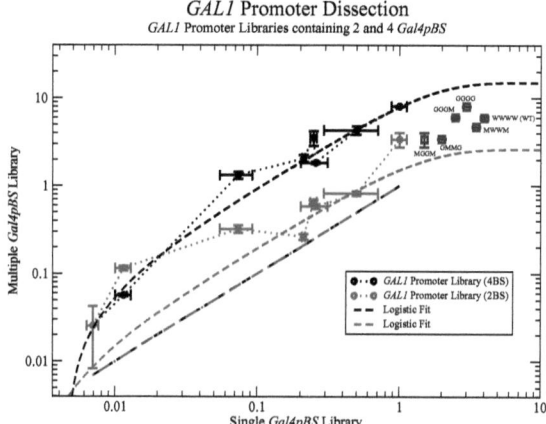

Figure 54: Multiple (four $Gal4pBS$) versus double $GAL1$ promoter libraries. The diagonal line represents the reference system (single versus single operator library). The black symbols represent the enrichment with respect to the single operator library of the multiple operator library consisting of four $Gal4pBS$ while the red symbols represent the double operator library. The blue symbols represent gene expression of additional constructs: M stands for mutated Gal4p binding site. G for Gal80 binding site. W for wild type binding sites. The position of the letter corresponds from left to right to distal to proximal binding sites positions in the promoter sequence. Hence, MGGM denotes $GAL1$ promoter with the first, distal binding site, mutated, followed by two Gal80 binding sites and a last proximal mutated binding site. MWWM consists of the first mutated binding site followed by the two native binding sites corresponding to the two central ones present in the wild type $GAL1$ promoter and at the end a last mutated binding sites. Similarly WWWW indicates the wild type promoter with binding sites according to native positions. Red and black dashed lines are non-linear curve fittings to the logistic equation (standard logistic function (Eq.59) where the resulting parameters are $A0 = 83.2863$; $A1 = 1.90373$; $A2 = 0.992807$; $A3 = -28.7271$ for the black curve and $A0 = 34.6288$; $A1 = 3.16372$; $A2 = 1.02073$; $A3 = -8.32232$ for the red curve) which have been subsequently used to compute logarithmic sensitivities represented in Fig.55. The experiments were obtained with strains and plasmids detailed in Tables 4, 5, 6, 7, 8.

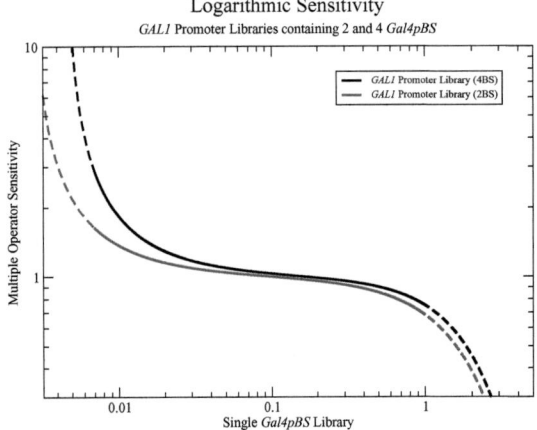

Figure 55: Sensitivity analysis of the curves of Fig.54. Continuous lines indicate experimentally accessible range, dotted lines are extrapolated based on the fitted function.

Part II
Transcriptional Interference: Interaction between Transcriptional Processes

1 Introduction to Transcriptional Interference

Transcriptional interference is mechanism of regulation that involves antagonistic relationship between transcription units. Different modes of transcriptional interference have been studied in the article hereby attached. Upstream and downstream interference are the two main topics in which I had a contribution. Cooperative effects at the level of transcriptional regulators binding represent the principal link with the work presented in 'Part I' of this thesis. These effects were studied here in the context of interacting transcriptional processes.

REPORT

Control and signal processing by transcriptional interference

Antoine Buetti-Dinh[1], Rosemarie Ungricht[1], János Z Kelemen, Chetak Shetty, Prasuna Ratna and Attila Becskei*

Institute of Molecular Biology, University of Zurich, Zurich, Switzerland
[1] These authors contributed equally to this work
* Corresponding author. Institute of Molecular Biology, University of Zurich, Winterthurerstrasse 190, Zurich 8057, Switzerland.
Tel.: +41 44 635 3180; Fax: +41 44 635 6811; E-mail: attila.becskei@molbio.uzh.ch

Received 28.11.08; accepted 21.7.09

A transcriptional activator can suppress gene expression by interfering with transcription initiated by another activator. Transcriptional interference has been increasingly recognized as a regulatory mechanism of gene expression. The signals received by the two antagonistically acting activators are combined by the polymerase trafficking along the DNA. We have designed a dual-control genetic system in yeast to explore this antagonism systematically. Antagonism by an upstream activator bears the hallmarks of competitive inhibition, whereas a downstream activator inhibits gene expression non-competitively. When gene expression is induced weakly, the antagonistic activator can have a positive effect and can even trigger paradoxical activation. Equilibrium and non-equilibrium models of transcription shed light on the mechanism by which interference converts signals, and reveals that self-antagonism of activators imitates the behavior of feed-forward loops. Indeed, a synthetic circuit generates a bell-shaped response, so that the induction of expression is limited to a narrow range of the input signal. The identification of conserved regulatory principles of interference will help to predict the transcriptional response of genes in their genomic context.
Molecular Systems Biology **5**: 300; published online 18 August 2009; doi:10.1038/msb.2009.61
Subject Categories: synthetic biology; chromatin & transcription
Keywords: noncoding transcription; promoter; repression

This is an open-access article distributed under the terms of the Creative Commons Attribution Licence, which permits distribution and reproduction in any medium, provided the original author and source are credited. Creation of derivative works is permitted but the resulting work may be distributed only under the same or similar licence to this one. This licence does not permit commercial exploitation without specific permission.

Introduction

One of the major goals of quantitative modeling of gene regulation is to predict gene expression based on the occupancy of gene regulatory sites by transcriptional factors. The action of transcriptional activators and repressors bound to a promoter can be represented as a mathematical operation. These operations have been systematically analyzed in prokaryotes (Buchler et al, 2003; Hermsen et al, 2006; Cox et al, 2007), and in eukaryotes (Ratna et al, 2009).

The above models focused on the classical role of transcriptional activators: the enhancement of gene expression. Interestingly, activators can also suppress gene expression by, at least, two different mechanisms (Shearwin et al, 2005).

First, intergenic transcription initiated by activators from upstream sequences can interfere with the expression of downstream genes. This upstream interference has been observed for the SER3, ADH1 and ADH3 genes in yeast (Martens et al, 2004, 2005; Bird et al, 2006). Intergenic transcription produces noncoding RNAs that have been detected in Saccharomyces cerevisiae and higher eukaryotes in large numbers (Hongay et al, 2006; Khaitovich et al, 2006; Neil et al, 2009; Xu et al, 2009). Positive regulatory aspects of transcriptional interference have been increasingly recognized in processes and phenomena, such as T-cell receptor recombination, latency of the HIV infection and epigenetic cellular memory (Schmitt et al, 2005; Abarrategui and Krangel, 2007; Lenasi et al, 2008).

Second, when an activator binds to a site that overlaps or is positioned downstream of the transcriptional initiation site, it can interfere with transcriptional initiation and elongation. This downstream antagonism is exemplified by the ZRT2, PRY3 and ACC1 genes (Li and Johnston, 2001; Bird et al, 2004; Bickel and Morris, 2006).

Signals passed onto transcriptional activators that either interfere with transcriptional initiation or initiate intergenic transcription are processed 'horizontally' along the DNA, which is mediated predominantly by the polymerase. Little is known about how these antagonistic signals are combined. Using the yeast, *Saccharomyces cerevisiae*, as a eukaryotic model organism, we have explored the principles of this signal conversion and how these signals can be utilized to control gene expression.

Results and discussion

Competitive inhibition by upstream interference

We studied upstream transcriptional interference using chromosomally integrated gene constructs. In these constructs, intergenic transcription interferes with the expression of a downstream *GFP* reporter gene under the control of different promoters (Figure 1, Materials and methods section). The intergenic transcription was triggered by the transcriptional activator GEV, activity of which was modulated by estradiol. First, we used the *ADH1* promoter, which has been already shown to be regulated by interference at its original genomic locus (Bird et al, 2006). On activation of GEV using estradiol, the GFP expression driven by the *ADH1* promoter decreased in a graded way, so that the expression had a unimodal distribution in the cell population (Figure 1A, Supplementary Figure S8). Thus, the mean expression level can be adequately used to quantify the output of the system.

Next, we measured the changes in the mean GFP expression as the occupancy of the activator-binding sites within a downstream promoter was varied. For this purpose, doxycycline was used to modulate the binding of the transcriptional activator, rtTA, to two *tet* operators within the downstream promoter. The doxycycline-induced binding of rtTA to the promoter led to GFP expression (Figure 1B). We observed that the suppression of GFP expression by intergenic transcription was gradually relieved as the rtTA binding strengthened, when

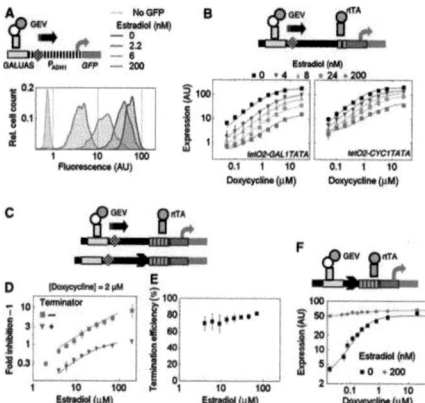

Figure 1 Upstream interference by intergenic transcription. (A) The 1000-bp long P_{ADH1} includes an upstream TATA box (−936 bp, checkered diamond), which is required to drive the intergenic transcription. Single cell distribution of GFP expression driven by GALUAS-$P_{ADH1(-1000\ to\ -1)}$ (YAntH44.6) is shown when intergenic transcription was activated at different estradiol concentrations. Relative cell count is shown. (B) The upstream activating sequence (UAS) of GAL1, GALUAS, was positioned upstream of the truncated EGT2 gene (EGT2 −115 to 509) to emulate intergenic transcription. The truncated gene comprises the core promoter with a TATA box (−115 to 0) and part of the ORF (1-509). The downstream promoter was obtained by fusing the [tetO]$_2$ to the GAL1 or CYC1 core promoters including a TATA box (YABH39.4 and −38.2). Expression was induced by doxycycline at different fixed concentrations of estradiol. Error bars represent s.d. values calculated from three experiments. The curves represent fits of the non-equilibrium model of upstream interference (see Supplementary Information), with α=3, k_{ON}=0.015 nM^{-1} min^{-1}; k_p=1 min^{-1}, η=49.6, k=0.041 min^{-1}, k_B=10 min^{-1}, m=33.4. A_{cri}=9.9 nM and K_{ind}=6.1 to account for the induction by doxycycline, and P=0, 1.5, 3.5, 8.6 and 71 for the respective estradiol concentrations. k_{OFF}=0.11 min^{-1}, β=2.8, bas=0.0055, v_{max}=201 for $P_{[tetO]2-GAL1TATA]}$; and k_{OFF}=0.07 min^{-1}, β=7, bas=-0.03, and v_{max}=230 for $P_{[tetO]2-CYC1TATA]}$. (C) Constructs used to measure termination efficiency. The arrowhead denotes the ACT1 transcriptional terminator. The downstream promoter was obtained by inserting five tet operators into the EGT2 promoter. (D) Expression was induced by 2 μM doxycycline and was inhibited by increasing concentrations of estradiol. Data are shown for GALUAS-EGT2 (−115 to 509)-$P_{[tetO]5inEGT2}$ (YAntH41.1, red squares) and GALUAS-EGT2 (−115 to 509)-T_{ACT1}-$P_{[tetO]5inEGT2}$ (YAntH42.1, blue triangles). (E) The termination efficiency was calculated from two independent experiments (see Materials and methods section) using the data as shown in (D). (F) Activation of gene expression by GEV when only the T_{ACT1} separates the GALUAS from $P_{[tetO]5inEGT2}$ (YAntH43.1). Source data is available for this figure at www.nature.com/msb.

the doxycycline concentration was increased from intermediate to high levels (Figures 1B and 2B). This indicates that intergenic transcription competes with the rtTA-driven transcription.

We compared the expression data obtained at different strengths of intergenic transcription with basic equilibrium models of repression (Box 1). The data agreed well with the model of competitive inhibition, at weak and moderate intergenic transcription (Figure 2A and B). The drop of inhibition efficiency close to saturation of activator-binding sites, a hallmark of competitive inhibition, was particularly pronounced with a promoter containing seven operators (Figure 2C and Supplementary Figure S9C and D), in which a higher degree of operator occupancy can be attained due to cooperative binding of rtTA (Becskei et al, 2005). The overwhelming majority of the data points aligned closely with the fitted curves calculated from the equilibrium competition model (Figure 2C).

The equilibrium competition model did not fit the data when interference was strong (Supplementary Figure S10). In this case, a good fit was obtained only to the data points measured at high concentration of doxycycline (Figure 2B and C). When these fits were extended into the range of low doxycycline concentrations, they overestimated the inhibition of gene expression, suggesting that intergenic transcription can have a positive effect on GFP expression (Figure 2B and C). To evaluate how general the above observations are, we studied interference when expression at the downstream promoter was driven by various activation domains (VP16 and Swi5) and by various core promoters (CYC1, GAL1 and EGT2). In all the examined cases, the two hallmarks were conserved, and only the overall efficiency of inhibition varied (Supplementary Figure S12); the competition dominated at medium and high doxycycline concentrations, whereas the positive effect of strong intergenic transcription was unmasked at a low doxycycline concentration.

Next, we built a detailed model on the basis of realistic molecular mechanisms. Previous studies have suggested that competition by intergenic transcription can arise when the elongating polymerase occludes the activator-binding sites (Sneppen et al, 2005). Furthermore, the elongating polymerase roadblocked by the activator can exert a force on the activator-DNA complex and destabilize it, after which the activator dissociates (Prescott and Proudfoot, 2002; Mosrin-Huaman et al, 2004; Galburt et al, 2007). However, when the polymerase traverses binding sites within the promoter, they can become more accessible, possibly due to changes in the chromatin structure, which enables the facilitated rebinding of transcription factors to them (Uhler et al, 2007). It is important to note that different methods for measuring DNA-protein interactions can produce contrasting results for the binding of the same transcription factor, when exposed to intergenic transcription (Bird et al, 2006). We constructed a non-

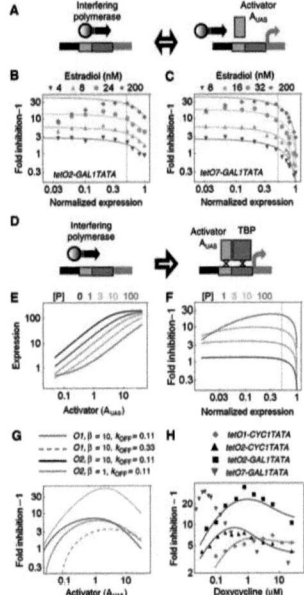

Figure 2 Equilibrium (A–C) and non-equilibrium competition models (D–H) of upstream interference. (A) Scheme of the equilibrium competition model. The downstream promoter is occupied either by the interfering polymerase or by the activator, A_{UAS}. (B, C) Equation (1) (Box 1) was fit to the data. $K_D^0=0.37$ and $f(R)=2.9, 5.9, 14.1^*$ and 41.4^* for $P_{tetO2\text{-}GAL1TATA}$ (data re-plotted from Figure 1B) (B); $K_D^0=0.024$ and $f(R)=2.6, 5.9, 18.5$ and 42.7^* for $P_{tetO7\text{-}GAL1TATA}$ (YABH34.5) (C). The asterisked $f(R)$ values were obtained by fitting equation (1) to data points that had a normalized expression higher than 0.4 (see Materials and methods section). (D) In the non-equilibrium competition model, the interfering polymerase traverses the UAS and the TATA box in the downstream promoter, after which they bind the activator, A_{UAS}, and the TBP with a higher affinity. (E) Gene expression as a function of A_{UAS} was calculated from the non-equilibrium model with the parameter values fitted for $P_{tetO2\text{-}GAL1TATA}$ (Figure 1B). The concentration of the activator [P] driving the intergenic transcription is color coded. (F) Curves were re-calculated from (E). (G) Fold inhibition at P=100 was calculated for promoters with one (O1) and two (O2) operators as in (F), except for the parameters specified in the figure legend. The red dashed line stands for one operator with reduced affinity. (H) Fold inhibition was measured at 200 nM estradiol as the doxycycline concentration was varied. The curves were fit with the parameter values obtained for the corresponding constructs in (Figure 1B). To link the A_{UAS} concentration to the doxycycline concentration, $A_{tot}=10.3$ nM and $K_{eq}=2.6$ were fit for promoters with CYC1TATA, measured on the same day. For $P_{tetO1\text{-}CYC1TATA}$ (YABH40.6), $k_{ON}=0.0072$ nM^{-1} min^{-1} and $k_{OFF}=0.13$ min^{-1} were fitted to account for its lower binding constant in comparison with $P_{tetO2\text{-}CYC1TATA}$. Source data is available for this figure at www.nature.com/msb.

Box 1 Competitive and non-competitive inhibition of gene expression

Two basic forms of inhibition of gene expression are described by simple equilibrium models. When an inhibitor, R, interferes with the binding of the transcriptional activator, A, inhibition of gene expression is competitive. Expression is given by

$$\mathrm{Ex} = w \frac{A}{K_D^A(1+f(R))+A} \quad (1)$$

K_D^A is the dissociation constant of the activator binding, w is a proportionality constant, whereas $f(R)$ is a lumped parameter incorporating the concentration and the dissociation constant of the inhibitor.

If the inhibitor does not prevent the activator from binding to the promoter, but suppresses transcription at a later stage, inhibition is non-competitive. A more general model incorporates the synergistic binding of the activator and the inhibitor, as well.

$$\mathrm{Ex} = w \frac{A}{K_D^A + A + K_D^A f(R) + \alpha A f(R)} \quad (2)$$

α denotes to what extent more likely is the joint binding of the activator and inhibitor than the binding assuming no interaction between them.

Characteristic profiles of inhibition, across a broad range of expression levels, can be conveniently compared when the fold change of expression due to a fixed concentration of the inhibitor is calculated as the activator concentration is varied. For this purpose, fold inhibition-1 was plotted against normalized expression (see Materials and methods section). Normalized expression, NE, corresponds to the expression, Ex, calculated in the absence of the inhibitor ($f(R)=0$, $w=1$ in equations (1) and (2)); NE=A·$(K_D^A+A)^{-1}$.

Competitive inhibition is shown for K_D^A=0.043, $f(R)$=3.8 in equation (1).

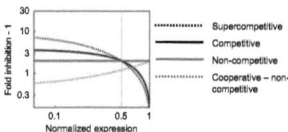

When gene expression approaches saturation, fold inhibition-1 drops rapidly, because the activator does not increase the expression noticeably but can increasingly outcompete R. However, fold inhibition-1 doubles at most, when the normalized expression is reduced from 0.5 to an arbitrary low value. For the supercompetitive mechanism, fold inhibition-1 increases more than twice when the normalized expression is reduced from 0.5 to an arbitrary low value. The supercompetitive curve is plotted for K_D^A=0.043, $f_A(R)$=0, $f_R(R)$=0.1 and α=0.011 using equation (4) given by Ratna et al, 2009. Supercompetitive inhibition arises when the activator and the repressor jointly determine the permissive state of the promoter (Ratna et al, 2009).

For non-competitive inhibition, fold inhibition-1 has a constant value as transcriptional activation is varied (the curve is shown for K_D^A=0.043, $f(R)$=2 and α=1 in equation (2)). When R and the activator bind cooperatively (K_D^A=0.043, $f(R)$=0.5 and α=4 in equation (2)), fold inhibition-1 decreases with decreasing transcriptional activation (also see Supplementary Figure S3).

equilibrium model that includes the rebinding of rtTA facilitated by a factor of β, and the increased association of the TATA-binding protein (TBP) and the above forms of competition (Figure 2D, Supplementary Figure S1).

The model successfully reproduced the positive effect on gene expression while preserving the hallmarks of competition (Figure 2E and F, Supplementary Figure S2). The model fitted to data obtained for promoters containing two tet operators (Figure 1B) also predicts that the peak value of inhibition by intergenic transcription does not increase when the interfering polymerase competes with the rtTA bound to only one operator, or may even decrease if the single operator has a lower affinity (Figure 2G). Inhibition of expression caused by strong intergenic transcription reaches its peak value at intermediate doxycycline concentrations, at which intergenic transcription can more easily outcompete rtTA-induced transcription and the positive effect is still negligible. The above prediction was consistent with the measurements: the peak inhibition for the tetO1-CYC1TATA construct was lower, and shifted to higher doxycycline levels in comparison with the tetO2-CYC1TATA construct (Figure 2H, Supplementary Figure S11). Similarly, when the number of tet operators, fused to a GAL1 core promoter, was increased from two to seven, the inhibition curve only shifted towards lower doxycycline values but its peak value did not change (Figure 2H).

Initiation and termination of transcriptional interference

The prevalence of interference in the genomic context depends on the number of DNA sequences that can initiate interference, and on how efficiently transcriptional terminators terminate transcription between adjacent genes to prevent interference. With some modifications, the above gene constructs can shed light on how likely interference arises at a given segment of the genome (Supplementary Figure S13).

When two different activators bind to a promoter, both of them can activate gene expression (Supplementary Figure S13C). Surprisingly, the insertion of a short, eight-nucleotide long TATA sequence converts the activator recruited to a site upstream of the TATA box into an inhibitor of expression (Supplementary Figure S13B). This inhibition, triggered by a simple sequence, was less efficient than the inhibition by intergenic transcription initiated by a full promoter (Supplementary Figure S13A, B and D).

The efficiency of termination was extrapolated by measuring to what extent interference was reduced when a terminator was inserted between the intergenic transcription unit and the downstream promoter–GFP constructs (Figure 1C). The efficiency of termination was calculated at intermediate rtTA binding, when the inhibition of expression is linearly dependent on estradiol concentration (Figure 1D). The ACT1 transcriptional terminator had a relatively constant, around 80%, termination efficiency over a broad range of intergenic transcription rates (Figure 1E). This constancy of the efficiency is surprising because it has been commonly assumed that terminators fail when the transcription rate passes a threshold value.

The genomes of yeast species are very compact and transcriptional terminators often overlap with the promoters of downstream genes (Valerius et al, 2002). Therefore, termination efficiency inferred from the changes in transcriptional interference is important to assess how efficiently terminators can isolate the transcriptional regulation of two adjacent genes.

Interestingly, the terminator did not reduce transcriptional activation when it was inserted directly downstream of a UAS that had no TATA box associated with it (Figure 1F). This suggests that terminators do not prevent an activator from initiating transcription at a TATA box positioned downstream of the terminator (Figure 1F), but they terminate transcription that had been fully initiated (Figure 1D).

Genome-wide analyses of gene expression suggest that genes may not reach their optimal expression level because of interference of their genomic environment (Liao and Zhang, 2008); and evolution proceeds in a direction to increase intergenic distances, whereby interference is reduced (Chiaromonte et al, 2003; Byrnes et al, 2006). Indeed, the simple DNA sequence requirements for the interference to be initiated (Supplementary Figure S13B) and, the good, but finite (~80%), efficiency of transcriptional termination between adjacent genes may contribute to the widespread occurrence of interference in the genome.

Downstream antagonism

To study antagonism by downstream activators, activator-binding sites were inserted downstream of the TATA box in the promoter–GFP constructs. Binding of GEV to the upstream site, GALUAS, drove the expression of GFP (Figure 3A). The binding of rtTA to *tet* operators downstream of a TATA box inhibited GFP expression (Figure 3A). This indicates that in addition to DNA-binding protein domains alone (Brent and Ptashne, 1984; Murphy et al, 2007), full-length transcriptional activators can interfere with the transcriptional activation. Expression data at different strengths of downstream antagonism were in excellent agreement with an equilibrium model for non-competitive inhibition that incorporates the cooperative binding of GEV and rtTA (Box 1, Figure 3A and B) (Cornish-Bowden, 2004). Cooperative binding of rtTA to promoters has been observed (Becskei et al, 2005). The cooperative interaction between the upstream and downstream sites could account for the observation that at a low estradiol concentra-

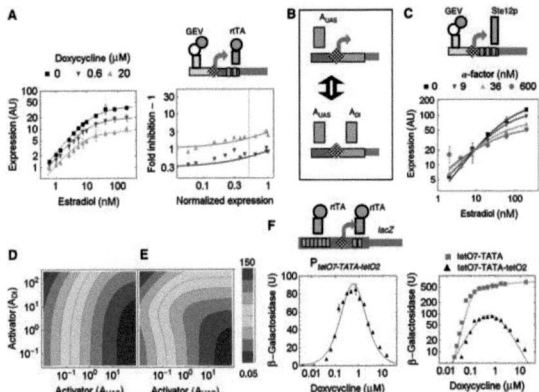

Figure 3 Downstream antagonism. The TATA box is denoted by a checkered diamond in the genetic constructs. Error bars represent s.d. values calculated from three experiments, unless otherwise specified. (A) Expression driven by $P_{GALUAS-TATA-tetO2}$ (RUY20) in the presence of different fixed concentrations of doxycycline. The curves were obtained by fitting equation (2) (Box 1): K_D^a=0.067, α=3.2. $f(R)$=0.31 and 1.05 for respective doxycycline concentrations (B) Scheme of non-competitive inhibition. When the A_{UAS} and A_{DI} activators bind to the promoter simultaneously, no transcription is initiated. (C) Expression driven by $P_{GALUAS-TATA-FUS1UAS}$ (YABH42.1) in the presence of different fixed concentrations of α-factor. The *FUS1*UAS contains three binding sites for the endogenous Ste12p transcriptional activator. Expression was adjusted using the $P_{GALUAS-TATA-MnFUS1UAS}$ construct (YABH43.2) to account for the nonspecific effects of α-factor on expression (see Materials and methods section). The curves are fits to the non-equilibrium model of the downstream antagonism (see Supplementary Information, SEq2) with p_{UAS}=0.01 nM^{-1} min^{-1}, p_{DI}=0.005 nM^{-1} min^{-1}, α=32.8, m_1=0.07 min^{-1}, m_2=0.1 min^{-1}, p_3=1 min^{-1}, m_3=0.2 min^{-1} and k=0.2 min^{-1}, A_{tot}=500 nM, K_{ind}=2161 nM, v_{max}=619; [A_{DI}]=0, 0.89, 2.79 and 4.46 for the respective α-factor concentrations. (D, E) Contour plots represent expression levels as a function of A_{DI} and A_{UAS} using the parameter values as in (C), except for p_{DI}=0.01 nM^{-1} min^{-1}, and the cooperativity of binding, α was varied: α=1 (D) and α=20 (E). (F) Expression driven by the $P_{tetO7-TATA-tetO2}$ (RUY65) and $P_{tetO7-CYC1TATA}$ (RUY67.13) constructs. LacZ was used to detect gene expression with higher sensitivity. Expression of the $P_{TATA-tetO2}$ construct (RUY69), which lacks an upstream activation sequence, is below the detection limit. The value for s.d. is calculated from two experiments. The bell-shaped curve was obtained by fitting SEq. 4 (w=470, n-m=1.2, N=0.46 and M=0.79). Source data is available for this figure at www.nature.com/msb.

tion, inhibition is weaker than predicted by a model of pure non-competitive inhibition.

To further explore the cooperative interaction between multiple binding events, three binding sites were inserted downstream of the TATA box, which are recognized by the Ste12p activator (Figure 3C). Ste12p activates its target genes when induced by α-factor, but it inhibits the expression of the *PRY3* gene by binding to a downstream site (Bickel and Morris, 2006). When the expression of the resulting $P_{GAL1UAS\text{-}TATA\text{-}FUS1UAS}$ construct was induced by high concentration of estradiol, addition of α-factor inhibited transcription in a dose-dependent manner (Figure 3C). However, at lower estradiol concentration, expression was paradoxically increased in response to increasing α-factor concentrations. This illustrates that interference in the same genetic construct can result in both inhibition and activation. Correspondingly, Ste12p buffers the action of GEV when induced by high concentration of α-factor, as expression changes only slightly over a broad range of estradiol concentrations.

The paradoxical activation can be explained when the above model was modified so that the activator bound to the downstream site induced a weak expression indirectly by a non-equilibrium kinetic effect on the upstream activator or directly by recruiting the transcription initiation machinery (Figures 3C and Supplementary Figures S4–S6).

Cooperative interactions are typically considered to be advantageous for regulation. For example, cooperative binding of repressors increases the sensitivity of response, making repression respond to environmental stimuli in a switch-like manner (Oehler *et al*, 2006). Thus, the reduction of inhibition by the cooperative interaction of the two antagonist activators seems rather disadvantageous for regulation. However, a two-dimensional input plot reveals that the cooperativity renders the response more square-like, so that high expression is restricted to a quadratic domain in which the occupancy of the upstream sites is high and that of the downstream sites is low (compare Figure 3D and E).

A genome-wide search retrieved many activator-binding sites downstream of a TATA box that are conserved in related yeast species (Supplementary Table S1). Some of these sites may regulate gene expression. A Mac1p-binding site downstream of the TATA box of the *FTR1* promoter inhibited gene expression, and activated expression when transferred to an upstream site (Supplementary Figure S15). This finding may explain in part why the deletion of Mac1p, a copper-responsive transcriptional activator, results in an increase in *FTR1* expression (De Freitas *et al*, 2004).

In all the retrieved promoters (Supplementary Table S1), a single binding site downstream of the TATA box was identified. Therefore, downstream antagonism is expected to follow the non-competitive inhibition with cooperative binding (Figure 3A).

Bell-shaped response

Both the equilibrium and non-equilibrium models predict that binding of the same activator to both the upstream and downstream sites generates a bell-shaped response (Supplementary Figure S7). When rtTA binds to *tet* operators flanking the TATA box, increasing doxycycline concentration resulted in a bell-shaped response: expression initially increased and after reaching a plateau, it declined (Figure 3F, Supplementary Figure S16). The peak expression in the bell-shaped response is around five times less than the maximal expression of the corresponding expression cassette containing the upstream activation sequence only, confirming the predictions of the model. As the bell-shaped response limits gene expression to a narrow range of inducer concentrations it has the ability to translate concentration gradients into localized expression patterns, similar to the stripe formation during embryonic development (Sanchez and Thieffry, 2003; Basu *et al*, 2005). This response is reminiscent of the output of gene circuits with feed-forward loops (Mangan and Alon, 2003; Kaplan *et al*, 2008).

The regulatory architecture of the *ZRT2* promoter is very similar to our construct that generated bell-shaped response (Figure 3F). The Zap1p activator binds to sites flanking the TATA box of the *ZRT2* promoter. Although prior experiments have focused on repression by Zap1p, the full data set is compatible with a bell-shaped response to zinc (Supplementary Figure S17) (Bird *et al*, 2004). This similarity underscores the utility of studying signal processing by synthetic circuits to understand the functioning of natural gene networks.

Conclusions

It is essential to know to what extent gene expression can be inhibited as transcriptional activation is varied to understand gene regulation and to design gene expression systems for biotechnological purposes. Simple equilibrium models were consistent with most of our observations. In particular non-competitive inhibition with cooperative binding is consistent with the findings on downstream antagonism, including the bell-shaped response, which has been observed in the genomic context as well (Bird *et al*, 2004). The equilibrium approach is frequently used to describe gene regulation when transcription factors bind to promoters, because binding is a reversible process and rapid relative to the kinetics of the reporter gene expression. A more complex non-equilibrium model is realistic to explain interference, for which the regulator is transcription itself. Transcription is a highly irreversible process as the elongating polymerase proceeds only in one direction. This energy consuming interfering polymerase regulates the expression of the gene it traverses. Therefore, the non-equilibrium approach can capture the irreversible nature of processes having a function in upstream interference. Interestingly, intergenic transcription inhibits the expression at $P_{tetO2\text{-}CYC1TATA}$ less efficiently than at $P_{tetO2\text{-}GAL1TATA}$, even though rtTA has nearly equal affinities for these promoters (Figure 2B). The fitted non-equilibrium model suggests that rtTA rebinds and restores the initiation complex more rapidly at $P_{tetO2\text{-}CYC1TATA}$, after the polymerase traverses the promoter. This may explain why the expression at $P_{tetO2\text{-}CYC1TATA}$ is more resistant to interference.

Our findings reveal unexpected links between different forms of transcriptional regulation. Both upstream interference and classical repression by repressor proteins in yeast rely on competitive inhibition even though they represent distinct molecular mechanisms (Ratna *et al*, 2009). These simple

Materials and methods

Genetic constructs

GEV is a fusion protein consisting of the Gal4p DNA binding domain, an estradiol receptor domain and the transcriptional activation domain, VP16-AD (Louvion et al, 1993). rtTA is a fusion of the rtetR DNA binding domain and the VP16-AD (Urlinger et al, 2000). The rtetR-(NLS-AD)^{8wt8} fusion was obtained by linking the following DNA sequences: rtetR(S2)(1–643), GGGCGCGCC, SWI5(1900–2125), CCTGCAGGG and SWI5(4–1639). rtTA(S2) served as a template for producing rtetR(S2). We used the enhanced green and the yellow variants of the green fluorescent protein as specified in Supplementary Table S2.

Yeast strains and growth conditions

All strains are congenic with W303 (ade2-1, leu2-3, ura3, trp1-1, his3-11,15 and can1-100). Genetic constructs were integrated into the chromosome (Supplementary Table S3).
Cells containing inducible gene expression constructs were grown for 5 h after induction in minimal medium, until a cell density of OD$_{600}$=0.4–0.8, unless specified otherwise, was attained. When α-factor was added to cell culture at OD$_{600}$=0.025, cells were grown for 200 min.
Co-expression of GEV and rtTA did not affect the growth rate of the cells and the percentage of cells that lost the constructs containing the reporter gene and rtTA was <0.01% (Supplementary Figure S14).

β-Galactosidase assay

β-Galactosidase activity was measured using cell extracts obtained from freeze-thaw cycles and CPRG was used as a substrate.

Flow cytometry and calculation of expression

Total fluorescence of, at least, 5000 cells was measured using flow cytometry. About 5–15% of total cell population was selected in the forward-scatter versus side-scatter plot to measure GFP fluorescence of cells with similar size. To calculate expression (Ex), the total fluorescence of GFP was divided by the background fluorescence of a control strain (F_C), which expresses lacZ only:

$$Ex_{e,d} = \frac{F_{e,d}}{F_C} - 1$$

The e and d subscripts refer to the applied concentration of estradiol and doxycycline, respectively.
The α-factor causes changes in the forward and side scatter of the cells in a concentration-dependent manner, and also in the expression level induced by estradiol. The latter effect of α-factor may be caused directly by changes in general transcriptional rates and/or indirectly caused by changes in cell growth and consequently in dilution rate of GFP. Therefore, expression was corrected by the expression of a construct in which the Ste12p-binding sites were mutated (YABH43.2):

$$Ex_{e,\alpha} = \frac{\left(\frac{FW_{e,\alpha}}{C_\alpha} - 1\right)\left(\frac{FM_{e,0}}{C_0} - 1\right)}{\frac{FM_{e,\alpha}}{C_\alpha} - 1}$$

where C_0 and C_α denote the control cell fluorescence at zero and the applied α-factor concentration, respectively. FW$_{e,\alpha}$ and FM$_{e,\alpha}$ denote GFP fluorescence of constructs with wild-type and mutant Ste12p binding sites, respectively. FM$_{e,0}$ corresponds to FM$_{e,\alpha}$ when α=0.
The background expression is typically low and is subject to large relative fluctuations when exposed to interference, which makes it difficult to discern the effect of interference when gene expression is not induced (Supplementary Table S4).

Data analysis

Normalized expression, NE, is the uninhibited expression at a given degree of activation divided by the maximally induced expression:

$$NE_{0,d} = \frac{Ex_{0,d}}{Ex_{0,d_{max}}} \text{ for upstream interference}$$

$$NE_{e,0} = \frac{Ex_{e,0}}{Ex_{e_{max},0}} \text{ for downstream antagonism}$$

For upstream interference d_{max}=20 μM; for downstream antagonism e_{max}=200 nM. Fold inhibition at a given point of normalized expression was obtained by dividing expression in the absence of antagonism by the expression suppressed by the antagonistic activator:

$$Fl_{e,d} = \frac{Ex_{0,d}}{Ex_{e,d}} \text{ for upstream interference}$$

$$Fl_{e,d} = \frac{Ex_{e,0}}{Ex_{e,d}} \text{ for downstream antagonism}$$

The characteristic profiles of weak inhibition on logarithmic plots are better displayed with fold inhibition-1 than with fold inhibition.
The termination efficiency, TE, was calculated by

$$TE = \left(1 - \frac{Fl_e(T) - 1}{Fl_e(NT) - 1}\right) \times 100$$

Fl_e denotes the fold inhibition of expression owing to the presence of estradiol. T and NT in the parentheses stand for the constructs with and without transcriptional terminator, respectively.

Model fitting

When the intergenic transcription was strong, the equilibrium competition model did not approximate well the full data set (Supplementary Figure S10), although it agreed well with the data obtained at high concentrations of doxycycline (Figure 2B and C). In such cases, the equation was fitted to only those data points that had a normalized expression higher than 0.4. This approach exposes how the remaining data points, obtained at lower doxycycline concentrations, deviate from the competition model. Subsequently, the competition model can be complemented parsimoniously.
To fit the non-equilibrium models to the experimental data, the concentration of the functionally active transcriptional activator, A_{UAS}, has to be extrapolated from the inducer (estradiol or doxycycline) concentration, ind, and the total activator concentration, A_{tot}.

$$A_{UAS} = A_{tot} \frac{ind}{K_{ind} + ind}$$

K_{ind} is a lumped equilibrium constant and represents the transport of the inducer across the cell membrane, and the binding of the inducer to the activator. Its fitted value depends also on the activity fluctuations of the inducer.

Supplementary information

Supplementary information is available at the Molecular Systems Biology website (www.nature.com/msb).

Acknowledgements

We thank Beatrice Blattmann and Simone Scherrer for technical help; Walter Schaffner, Bernhard Dichtl and Denise Hengartner (BUSS) for helpful discussions; and Reine Byun for reading the paper. This study was supported by the Swiss National Foundation and the UZH-URPP.

Conflict of interest

The authors declare that they have no conflict of interest.

References

Abarrategui I, Krangel MS (2007) Noncoding transcription controls downstream promoters to regulate T-cell receptor alpha recombination. *EMBO J* **26:** 4380–4390
Basu S, Gerchman Y, Collins CH, Arnold FH, Weiss R (2005) A synthetic multicellular system for programmed pattern formation. *Nature* **434:** 1130–1134
Becskei A, Kaufmann BB, van Oudenaarden A (2005) Contributions of low molecule number and chromosomal positioning to stochastic gene expression. *Nat Genet* **37:** 937–944
Bickel KS, Morris DR (2006) Role of the transcription activator Ste12p as a repressor of PRY3 expression. *Mol Cell Biol* **26:** 7901–7912
Bird AJ, Biankman E, Stillman DJ, Eide DJ, Winge DR (2004) The Zap1 transcriptional activator also acts as a repressor by binding downstream of the TATA box in ZRT2. *EMBO J* **23:** 1123–1132
Bird AJ, Gordon M, Eide DJ, Winge DR (2006) Repression of ADH1 and ADH3 during zinc deficiency by Zap1-induced intergenic RNA transcripts. *EMBO J* **25:** 5726–5734
Brent R, Ptashne M (1984) A bacterial repressor protein or a yeast transcriptional terminator can block upstream activation of a yeast gene. *Nature* **312:** 612–615
Buchler NE, Gerland U, Hwa T (2003) On schemes of combinatorial transcription logic. *Proc Natl Acad Sci USA* **100:** 5136–5141
Byrnes JK, Morris GP, Li WH (2006) Reorganization of adjacent gene relationships in yeast genomes by whole-genome duplication and gene deletion. *Mol Biol Evol* **23:** 1136–1143
Chiaromonte F, Miller W, Bouhassira EE (2003) Gene length and proximity to neighbors affect genome-wide expression levels. *Genome Res* **13:** 2602–2608
Cornish-Bowden A (2004) *Fundamentals of Enzyme Kinetics*. London: Portland Press
Cox III RS, Surette MG, Elowitz MB (2007) Programming gene expression with combinatorial promoters. *Mol Syst Biol* **3:** 145
De Freitas JM, Kim JH, Poynton H, Su Y, Wintz H, Fox T, Holman P, Loguinov A, Keles S, van der Laan M, Vulpe C (2004) Exploratory and confirmatory gene expression profiling of macIDelta. *J Biol Chem* **279:** 4450–4458
Galburt EA, Grill SW, Wiedmann A, Lubkowska L, Choy J, Nogales E, Kashlev M, Bustamante C (2007) Backtracking determines the force sensitivity of II RNAP in a factor-dependent manner. *Nature* **446:** 820–823
Hermsen R, Tans S, ten Wolde PR (2006) Transcriptional regulation by competing transcription factor modules. *PLoS Comput Biol* **2:** e164
Hongay CF, Grisafi PL, Galitski T, Fink GR (2006) Antisense transcription controls cell fate in *Saccharomyces cerevisiae*. *Cell* **127:** 735–745
Kaplan S, Bren A, Dekel E, Alon U (2008) The incoherent feed-forward loop can generate non-monotonic input functions for genes. *Mol Syst Biol* **4:** 203
Khaitovich P, Kelso J, Franz H, Visagie J, Giger T, Joerchel S, Petzold E, Green RE, Lachmann M, Paabo S (2006) Functionality of intergenic transcription: an evolutionary comparison. *PLoS Genet* **2:** e171
Lenasi T, Contreras X, Peterlin BM (2008) Transcriptional interference antagonizes proviral gene expression to promote HIV latency. *Cell Host Microbe* **4:** 123–133
Li Q, Johnston SA (2001) Are all DNA binding and transcription regulation by an activator physiologically relevant? *Mol Cell Biol* **21:** 2467–2474

Liao BY, Zhang J (2008) Coexpression of linked genes in mammalian genomes is generally disadvantageous. *Mol Biol Evol* **25:** 1555–1565
Louvion JF, Havaux-Copf B, Picard D (1993) Fusion of GAL4-VP16 to a steroid-binding domain provides a tool for gratuitous induction of galactose-responsive genes in yeast. *Gene* **131:** 129–134
Mangan S, Alon U (2003) Structure and function of the feed-forward loop network motif. *Proc Natl Acad Sci USA* **100:** 11980–11985
Martens JA, Laprade L, Winston F (2004) Intergenic transcription is required to repress the *Saccharomyces cerevisiae* SER3 gene. *Nature* **429:** 571–574
Martens JA, Wu PY, Winston F (2005) Regulation of an intergenic transcript controls adjacent gene transcription in *Saccharomyces cerevisiae*. *Genes Dev* **19:** 2695–2704
Mosrin-Huaman C, Turnbough Jr CL, Rahmouni AR (2004) Translocation of *Escherichia coli* RNA polymerase against a protein roadblock *in vivo* highlights a passive sliding mechanism for transcript elongation. *Mol Microbiol* **51:** 1471–1481
Murphy KF, Balazsi G, Collins JJ (2007) Combinatorial promoter design for engineering noisy gene expression. *Proc Natl Acad Sci USA* **104:** 12726–12731
Neil H, Malabat C, d'Aubenton-Carafa Y, Xu Z, Steinmetz LM, Jacquier A (2009) Widespread bidirectional promoters are the major source of cryptic transcripts in yeast. *Nature* **457:** 1038–1042
Oehler S, Alberti S, Muller-Hill B (2006) Induction of the lac promoter in the absence of DNA loops and the stoichiometry of induction. *Nucleic Acids Res* **34:** 606–612
Prescott EM, Proudfoot NJ (2002) Transcriptional collision between convergent genes in budding yeast. *Proc Natl Acad Sci USA* **99:** 8796–8801
Ratna P, Scherrer S, Fleischli C, Becskei A (2009) Synergy of repression and silencing gradients along the chromosome. *J Mol Biol* **387:** 826–839
Sanchez L, Thieffry D (2003) Segmenting the fly embryo: a logical analysis of the pair-rule cross-regulatory module. *J Theor Biol* **224:** 517–537
Schmitt S, Prestel M, Paro R (2005) Intergenic transcription through a polycomb group response element counteracts silencing. *Genes Dev* **19:** 697–708
Shearwin KE, Callen BP, Egan JB (2005) Transcriptional interference—a crash course. *Trends Genet* **21:** 339–345
Sneppen K, Dodd IB, Shearwin KE, Palmer AC, Schubert RA, Callen BP, Egan JB (2005) A mathematical model for transcriptional interference by RNA polymerase traffic in *Escherichia coli*. *J Mol Biol* **346:** 399–409
Uhler JP, Hertel C, Svejstrup JQ (2007) A role for nonccoding transcription in activation of the yeast *PHO5* gene. *Proc Natl Acad Sci USA* **104:** 8011–8016
Urlinger S, Baron U, Thellmann M, Hasan MT, Bujard H, Hillen W (2000) Exploring the sequence space for tetracycline-dependent transcriptional activators: novel mutations yield expanded range and sensitivity. *Proc Natl Acad Sci USA* **97:** 7963–7968
Valerius O, Brendel C, Duvel K, Braus GH (2002) Multiple factors prevent transcriptional interference at the yeast ARO4-HIS7 locus. *J Biol Chem* **277:** 21440–21445
Xu Z, Wei W, Gagneur J, Perocchi F, Clauder-Munster S, Camblong J, Guffanti E, Stutz F, Huber W, Steinmetz LM (2009) Bidirectional promoters generate pervasive transcription in yeast. *Nature* **457:** 1033–1037

Molecular Systems Biology is an open-access journal published by *European Molecular Biology Organization* and *Nature Publishing Group*. This article is licensed under a Creative Commons Attribution-Noncommercial-Share Alike 3.0 Licence.

SUPPLEMENTARY INFORMATION

I. NON-EQUILIBRIUM MODEL OF UPSTREAM INTERFERENCE 2
    Figure S1. Scheme of processes in upstream interference. 2
    Figure S2. Solutions of the non-equilibrium model of upstream interference. 4

II. EQUILIBRIUM MODELS OF DOWNSTREAM ANTAGONISM 4
    Figure S3. Inhibition due to competitive and non-competitive mechanisms. 5
    Figure S4. Equilibrium model of paradoxical activation. 5

III. NON-EQUILIBRIUM MODEL OF DOWNSTREAM ANTAGONISM 6
    Figure S5. Scheme of processes in downstream antagonism. 6
    Figure S6. Paradoxical activation is reproduced by the non-equilibrium model. 7

IV. BELL-SHAPED RESPONSE 8
    Figure S7. Bell-shaped response generated by different models of downstream antagonism. 9

V. PARAMETER VALUES 10

FURTHER SUPPLEMENTARY FIGURES 11
    Figure S8. Effect of upstream interference on the ADH1 promoter. 11
    Figure S9. The effect of intergenic transcription on downstream promoters containing one or seven tet operators. 12
    Figure S10. Fitting of the equilibrium competition model to upstream interference. 13
    Figure S11. Induction curves for tetO1- and tetO2-CYC1TATA 13
    Figure S12. Conserved features in the response to intergenic transcription. 14
    Figure S13. Initiation of upstream interference. 15
    Figure S14. The effect of GEV and rtTA on cellular growth and on disintegration of chromosomally integrated gene constructs. 16
    Figure S15. The effect of activator binding sites downstream of TATA boxes on gene expression 18
    Figure S16. Bell-shaped response generated by the $P_{tetO7\text{-}TATA\text{-}tetO2}$-GFP construct. 19
    Figure S17. Bell-shaped response of the ZRT2 gene to a gradient of zinc concentration. 19

SUPPLEMENTARY TABLES 20
    Table S1. Conserved activator recognition sequences downstream of conserved TATA boxes. 20
    Table S2. Strain list. 22
    Table S3. Plasmid list. 26
    Table S4. Effect of antagonistic activators on the background gene expression. 29

SEQUENCES USED FOR THE GENE CONSTRUCTS 30

SUPPLEMENTARY REFERENCES 34

I. Non-equilibrium model of upstream interference

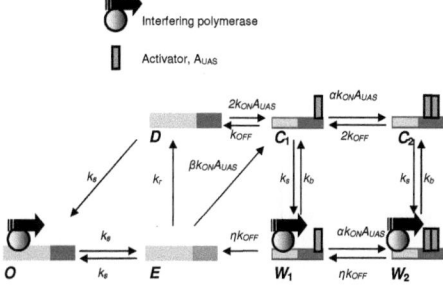

Figure S1. Scheme of processes in upstream interference. The configurations of the downstream promoter are denoted by bold capitals. The red and blue parts of the promoter symbolize the UAS with native and elevated affinity, respectively.

$$\frac{dO}{dt} = k_s \frac{P}{m+P}(D+E) - k_s O$$

$$\frac{dC_1}{dt} = 2k_{ON}A_{UAS}D + 2k_{OFF}C_2 + k_b W_1 + \beta k_{ON}A_{UAS}E - (k_{OFF} + k_s \frac{P}{m+P})C_1 - \alpha k_{ON}A_{UAS}C_1$$

$$\frac{dC_2}{dt} = \alpha k_{ON}A_{UAS}C_1 + k_b W_2 - (2k_{OFF} + k_s \frac{P}{m+P})C_2$$

$$\frac{dW_1}{dt} = k_s \frac{P}{m+P}C_1 + \eta k_{OFF}W_2 - (k_b + \eta k_{OFF})W_1$$

$$\frac{dW_2}{dt} = k_s \frac{P}{m+P}C_2 - (k_b + \eta k_{OFF})W_2$$

$$\frac{dE}{dt} = k_s O + \eta k_{OFF}W_1 - (k_r + \beta k_{ON}A_{UAS} + k_s \frac{P}{m+P})E$$

$$C_1 + C_2 + W_1 + W_2 + E + O + D = 1$$

151

The model describes the binding of the activator, A$_{UAS}$, to two operators within the upstream activating sequence (UAS) of the downstream promoter. The downstream promoter is exposed to an interfering polymerase, which is initiated by an activator, P, bound to the upstream promoter. The half-maximal activity of the upstream promoter is reached when the non-dimensional concentration of P is equal to m.
The following downstream promoter configurations were considered: D is the free downstream promoter, while O, C_1, and C_2 denote the downstream promoter occupied by the interfering polymerase, one or two molecules of A$_{UAS}$, respectively. W_1 and W_2 represent promoter configurations in which the interfering polymerase is stalled upstream of the bound A$_{UAS}$.
A$_{UAS}$ associates with the DNA by k_{ON} and dissociates by k_{OFF}. The binding rate of a second A$_{UAS}$ is facilitated by a factor of α, if one A$_{UAS}$ molecule is already bound. The interfering polymerase has a constant rate of progression in and out of the UAS, k_S. The UAS traversed by the polymerase, E, associates with the A$_{UAS}$, at a higher rate, $\beta\, k_{ON}$, $\beta > 1$. The stalled polymerase can dissociate from the DNA at a rate of k_b or it can enhance the dissociation rate of the A$_{UAS}$ bound to the promoter, by a factor of η.
Transcription is initiated when at least one A$_{UAS}$ molecule is bound to the promoter (C_1, C_2, W_1 and W_2). It is assumed, that after the polymerase traverses the TATA-box, it becomes highly accessible, possibly due to a destabilization of the nucleosome. This accessible TATA-box can bind the TATA binding protein (TBP), which leads to the initiation of transcription at a low level (*bas*). The occurrence of this event is proportional to the occupancy of the downstream promoter by the interfering polymerase (*O*). Thus, total gene expression of the reporter gene is given by:

$$v = v_{\max}(C_1 + C_2 + W_1 + W_2 + basO)$$

The steady-state solution of the respective differentiation equations can be solved explicitly, and yields a unique solution containing a long algebraic expression (not shown).
A more detailed model distinguishes two variants of the C1 and W1 configurations, depending on which one of the two binding sites is occupied. The model solutions are very similar to that one of the above simpler version.
The model describing a downstream promoter with a single operator can be obtained using above equations by substituting $\alpha = 0$, and dividing k_{ON} by two.

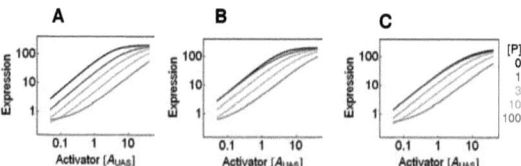

Figure S2. Solutions of the non-equilibrium model of upstream interference.
Expression of the reporter gene at different fixed intensities of intergenic transcription. Intergenic transcription is a hyperbolic function of the activator [P], bound to an upstream promoter. The curves were calculated from the algebraic solution of the model with $\alpha = 3$, $k_{ON} = 0.015$ nM^{-1}min^{-1}, $k_{OFF} = 0.11$ min^{-1}, $k_b = 1$ min^{-1}, $\eta = 49.6$, $k_r = 0.041$ min^{-1}, $k_S = 10$ min^{-1}, $m = 33.4$ and $bas = 0.0055$. The downstream promoter contains two (A and B) or one (C) binding sites for the activator. The rebinding efficiency of A$_{UAS}$ is $\beta = 2.8$ (A) and $\beta = 7$ (B and C).

II. Equilibrium models of downstream antagonism

The Eq. 2 (Box1) is a general equation and depending on the value of α, it is named as competitive (α = 0), non-competitive (α = 1) and mixed inhibition (for all other positive values of α) in enzyme kinetics. Within the context of promoter regulation it is intuitive to name non-competitive inhibition with cooperative binding, when α > 1. Figure S3 shows the characteristics of inhibition for different values of α.

Phenomena in interference and antagonism of transcriptional activators bear often more similarity to receptor theory in pharmacology (Kenakin, 2004). For example, partial agonists have a positive effect on the receptor response but can outcompete a more efficacious agonist, resulting in the net decrease of the response. The downstream activator Ste12p bears some resemblance to partial agonist when described by the following equation:

$$Ex = w \frac{A + e_d K_D^A f(R)}{K_D^A + A + K_D^A f(R) + \alpha A f(R)} \qquad \text{SEq. 1}$$

$f(R)$ is a lumped parameter incorporating the concentration and the dissociation constant of the downstream activator, while e_d is a proportionality constant for the expression induced by the downstream activator. In the above case, the downstream activator recruits the TBP (TATA-binding protein) and the pre-initiation complex with a low efficacy. This is sufficient to induce paradoxical activation (Figure S4). In the absence of GEV, a small degree of paradoxical activation is observed for Ste12p (Table S4). It is difficult to determine the contributions of the equilibrium and non-equilibrium mechanisms (SEq 1, 2) because the background expression is small and subject to large fluctuations.

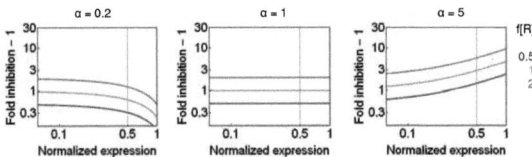

Figure S3. Inhibition due to competitive and non-competitive mechanisms.
Solution of Eq.2. (Box 1) is shown for various values of α. Decreasing values of α implicate an increasing contribution of competitive inhibition. $K_D^A = 0.043$.

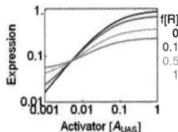

Figure S4. Equilibrium model of paradoxical activation.
Solution of SEq.1 is shown for various intensities of downstream activator binding, f(R), with the following parameter values $K_D^A = 0.06$, $e_d = 0.1$ and α = 3.

III. Non-equilibrium model of downstream antagonism

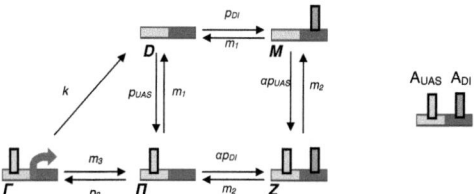

Figure S5. Scheme of processes in downstream antagonism.
Schematic representation of promoter configurations in the downstream antagonism constructs. The promoter configurations are denoted by bold capitals. The yellow and the blue rectangles represent the activators bound to the upstream and downstream sites, respectively.

$$\frac{dD}{dt} = m_1(\Pi + M) + k\Gamma - p_{UAS}A_{UAS}D - p_{DI}A_{DI}D$$

$$\frac{dM}{dt} = p_{DI}A_{DI}D + m_2Z - m_1M - \alpha p_{UAS}A_{UAS}D$$

$$\frac{d\Pi}{dt} = p_{AUS}A_{UAS}D + m_2Z + m_3\Gamma - (m_1 + p_3)\Pi - \alpha p_{DI}A_{DI}\Pi$$

$$\frac{d\Gamma}{dt} = p_3\Pi - (k + m_3)\Gamma$$

$$D + M + \Pi + Z + \Gamma = 1$$

$$v = v_{max}(m_3 + k)\Gamma$$

Hence,

$$v = v_{max}\frac{p_{UAS}(m_3 + k)m_2 p_3 A_{UAS}(2m_1 + \alpha p_{UAS}A_{UAS} + \alpha p_{DI}A_{DI})}{denom_1 + denom_2 + denom_3} \qquad \text{SEq. 2}$$

$$denom_1 = 2m_1^2 m_2 p_3(m_3 + k) + \alpha p_{UAS}p_{DI}A_{UAS}A_{DI}(kp_3 + (m_3 + k)\alpha(p_{UAS}A_{UAS} + p_{DI}A_{DI}))$$

$$denom_2 = m_2(2p_3kp_{DI}A_{DI} + (k + m_3)\alpha(p_{UAS}A_{UAS} + p_{DI}A_{DI})^2 + p_3p_{DI}A_{DI}\alpha(p_4 + p_{DI}A_{DI} + p_{DI}A_{DI}))$$

$$denom_3 = m_1[2(k+m_3)\alpha p_{UAS} p_{DI} A_{UAS} A_{DI} + m_2(2p_3(k+p_{UAS}A_{UAS}) + (2+\alpha)(k+m_3)(p_{UAS}A_{UAS} + p_{DI}A_{UAS}))]$$

The non-equilibrium model contains binding events of activators to the promoter that are equivalent to that in the equilibrium model. In addition, it is assumed that the polymerase binds to the core promoter when only the upstream activator is bound, leading to the configuration Γ. The polymerase and the activator can dissociate from the promoter simultaneously, a step which occurs irreversibly at a rate, k. p_{DI} and p_{UAS} denote the association rates for A_{DI} and A_{UAS} respectively. α is the cooperative increase in the association rate of the activator, when the other activator is already bound to the promoter. m_1 and m_2 stand for the dissociation rates of an activator, when one (configurations M with A_{DI} and Π with A_{UAS}) or two (configuration Z) activators are bound to DNA, respectively. D stands for the empty configuration. The above simplifications also take into account the principle of detailed balance in reversible reaction cycles. p_3 and m_3 stand for the rates of the association and dissociation of the polymerase, respectively. The amount of the polymerase is not limiting. Transcription occurs when the configuration Γ changes into another configuration.

SEq. 2 is the steady-state solution for the reporter gene expression derived from the above differential equations.

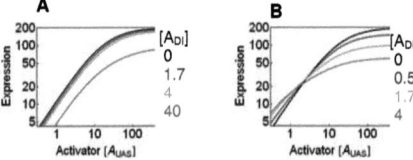

Figure S6. Paradoxical activation is reproduced by the non-equilibrium model. The same parameters values were taken for the model of downstream interference as in Figure 3C. $p_{UAS} = p_{DI} = 0.01$ nM^{-1}min^{-1}, $m_1 = 0.07$ min^{-1}, $m_2 = 0.1$ min^{-1}, $p_3 = 1$ min^{-1}, $m_3 = 0.2$ min^{-1} and $k = 0.2$ min^{-1}. (A) $\alpha = 1$ (cooperativity is absent). (B) $\alpha = 20$.

IV. Bell-shaped response

Next, we analyzed the equilibrium and non-equilibrium models to derive a response of a promoter with downstream antagonism, when the same activator binds to the upstream and downstream sites (Activator = $A_{UAS} = A_{DI}$).

(A) Equilibrium model:
The equilibrium model of non-competitive inhibition (Eq. 1 in Box1) can be rearranged in the following form when $\alpha = 1$:

$$Ex = w \frac{A}{K_A + A} \frac{1}{1 + f(R)} \qquad \text{SEq. 3}$$

When the same activator binds to the upstream and downstream sites ($A = f(R) = x$) with distinct degrees of cooperative binding and affinity, SEq. 3 can be generalized with the help of a Hill function:

$$Ex = w \frac{x^n}{N + x^n} \frac{M}{M + x^m} \qquad \text{SEq. 4}$$

The Hill-coefficients of binding to the upstream and downstream sites are denoted by n and m, respectively. The corresponding equilibrium dissociation constants are denoted by N and M.

Minimal requirements for the generation of a bell-shaped response are met if the expression level approaches zero at zero and very high induction, and if there is a single maximum at intermediate induction levels. In SEq 4., $Ex(x) = 0$ when $x \to 0$ and $x \to \infty$, while for $0 < x < \infty$, $Ex(x) > 0$.

To determine the number of maxima, the first derivative of the function is equated with

$$Ex'(x) = wM \frac{x^{n-1}}{(N + x^n)^2} \frac{NMn + (n - m)Nx^m - mx^{m+n}}{(M + x^m)^2}$$

Consequently,

$$NMn + (n - m)Nx^m - mx^{m+n} = 0$$

According to Descartes' rule of signs, the number of positive roots of this polynomial does not exceed 1. In combination with Rolle's theorem, which asserts that this function has at least one maximum in this region, it is evident that the function has a single maximum $0 < x < \infty$ for all positive real valued N, M, n and m.

Thus, a bell-shaped response is a general and robust property of the above equation. The above derivation does not addresses how peaked the response is; and for a broad range of parameter values, the response is relatively flat.

(B) Non-equilibrium model:

In order to obtain the bell-shaped response, the upstream activator, A_{UAS} was equated with the downstream interfering activator A_{DI} in SEq. 2.

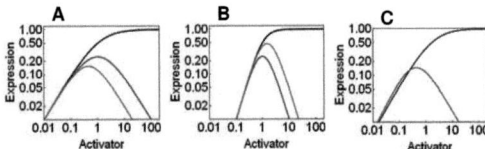

Figure S7. Bell-shaped response generated by different models of downstream antagonism. The black curves denote expression when the activator binds to the upstream sites only (**A**) Equilibrium model (Eq. 2 in Box1). $K_A = 1$; $\alpha = 1$ (blue lines) and $\alpha = 5$ (red lines). (**B**) Equilibrium model based on the Hill-function. SEq4; $N = 1$, $n = 2$, $m = 1$; while $M = 1$ (blue lines) and $M = 5$ (red lines). (**C**) Non-equilibrium model (SEq.2); $p_{UAS} = p_{DI} = 0.1$ nM^{-1}min^{-1}, $\alpha = 20$, $m_1 = 0.07$ min^{-1}, $m_2 = 0.1$ min^{-1}, $p_3 = 1$ min^{-1}, $m_3 = 0.2$ min^{-1} and $k = 0.2$ min^{-1}.

V. Parameter values

In order to fit the parameters of the non-equilibrium models, the parameter values were constrained to a realistic range. The range of values was set by comparing the relevant binding constants for three different transcriptional factors: the TATA-binding protein (TBP), the zinc-responsive transcriptional activator, Zap1p, and the regulator of carbon source utilization, Adr1p. In particular, we considered the on-rates [1 $M^{-1}s^{-1}$ = 6 10^{-8} $nM^{-1}min^{-1}$], the off-rates [1 s^{-1} = 60 min^{-1}] and the equilibrium dissociation constants, K_D [nM].
k_{ON} = 0.006, 4 and 0.012 $nM^{-1}min^{-1}$; k_{OFF} = 0.018, 2 and 0.8 min^{-1}; K_D = 3, 0.5 and 66 nM for TBP, Zap1p and Adr1p, respectively (Evans-Galea et al., 2003; Schaufler and Klevit, 2003; Wolner and Gralla, 2001).
Thus, the values of the parameters were constrained to the following ranges: k_{ON} = 0.005 to 5; k_{OFF} = 0.01 to 1, K_D = 0.1 to 100 nM.

For example, the following values were used for upstream interference:
k_{ON} = 0.015 $nM^{-1}min^{-1}$, k_{OFF} = 0.11 min^{-1}. Then, K_D = 7.3 nM.

Similarly for downstream antagonism:
$k_{ON} = p_{UAS} = p_{DI} = 0.1$ $nM^{-1}min^{-1}$; $k_{OFF} = m_1 = 0.07$ min^{-1}, $m_2 = 0.1$. Then, K_D = 0.7 and 1 nM.

Further supplementary figures

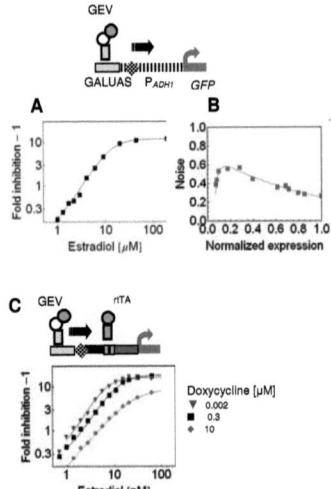

Figure S8. Effect of upstream interference on the *ADH1* promoter.
Comparison of the responses of the *ADH1* promoter to intergenic transcription with that of a synthetic promoter. The responses are similar regarding the magnitude and the functional form. (**A** and **B**) Expression is driven by *GALUAS*-P$_{ADH1}$ (YAntH44.6). (A). Noise equals the coefficient of variation of the fluorescence distribution (B).
(**C**) Expression driven by P$_{[tetO]2\text{-}SIC1TATA}$ (YABH50.31) at three different doxycycline concentrations.

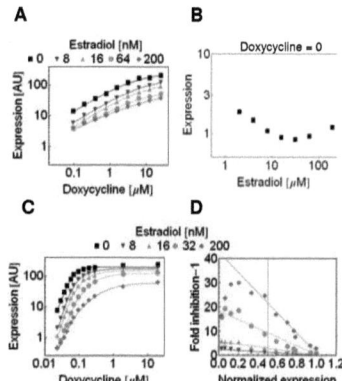

Figure S9. The effect of intergenic transcription on downstream promoters containing one or seven *tet* operators. Expression was induced by doxycycline at different fixed concentrations of estradiol. (**A** and **B**) $P_{tetO1\text{-}CYC1TATA}$ (YABH40.6). When expression is driven only due to a residual binding of rtTA, in the absence of doxycyline, the positive effect of intergenic transcription becomes apparent. As estradiol concentration increases, expression initially decreases and after reaching a minimum, expression rebounds (B). (**C** and **D**) $P_{tetO7\text{-}GAL1TATA}$ (YABH34.5). Data shown in (C) are re-plotted in (D) using a linear scale, and in Figure 2C using a logarithmic scale.

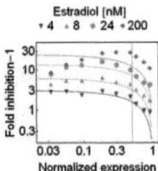

Figure S10. Fitting of the equilibrium competition model. Eq.1 (Box 1) was fit to the full data set obtained for $P_{tetO2\text{-}GAL1TATA}$: $K_D^A = 0.37$ and $f(R) = 2.9, 5.9, 13.7$ and 24.1 (in the order of increasing estradiol concentration). The data are identical with those shown in Figure 2B.

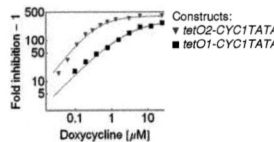

Figure S11. Induction curves for tetO1- and tetO2-CYC1TATA
A Hill function as defined within SEq. 4 was fit to the data: $N = 2.66$ $v_{max} = 307$ and $n = 0.87$ for $P_{tetO1\text{-}CYC1TATA}$; $N = 0.33$ $v_{max} = 404.9$ and $n = 1.14$ for $P_{tetO2\text{-}CYC1TATA}$. The difference in the affinities of rtTA to tetO1 and tetO2 is bigger than two, the expected value assuming that the affinities of the tet operators are independent of their sequence context. Therefore, it is likely that sequence context of the operators affects the affinity of the operators. Similar relations were observed for simple promoter constructs with varying numbers of *tet* operators (Becskei et al., 2005).

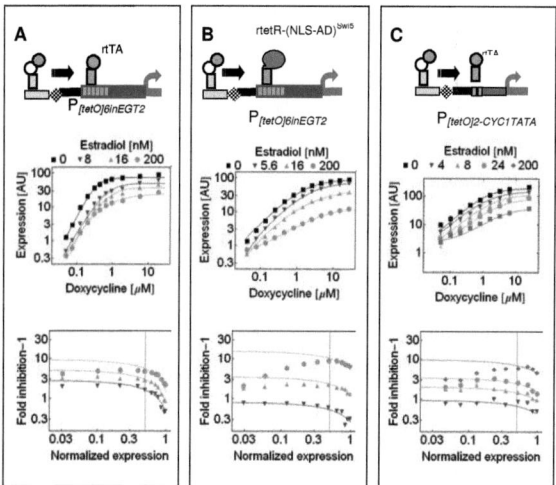

Figure S12. Conserved features in the response to intergenic transcription. The hallmarks of the upstream interference, competition and positive effect on gene expression, are conserved when different downstream constructs were exposed to intergenic transcription (first row). The curves in the fold inhibition-1 versus normalized expression plots represent fits of the equilibrium competition model (Eq.1 in Box 1). The asterisked $f(R)$ values were obtained by fitting Eq.1 to data points that had a normalized expression higher than 0.4 (see Materials and Methods). The gridline denotes half-maximal expression.
(**A**) Expression at P$_{[tetO]6inEGT2}$ driven by rtTA (YAntH16.1). P$_{[tetO]6inEGT2}$ is obtained by replacing the six Swi5p binding sites within the *EGT2* promoter by *tet* operators. $K_D^A = 0.18$ and $f(R) = 2.8^*, 5.1^*,$ and 9.5^*, for the respective estradiol concentrations.
(**B**) Expression at P$_{[tetO]6inEGT2}$ was driven by rtetR-(NLS-AD)^{Swi5} (YAntH17.13). $K_D^A = 0.47$ and $f(R) = 0.8, 3.5^*,$ and 15.7^*, for the respective estradiol concentrations.
(**C**) The data are taken from Figure 1B (YABH-38.2). $K_D^A = 0.8$ and $f(R) = 0.93, 2, 3.4^*,$ and 9.8^*, for the respective estradiol concentrations.

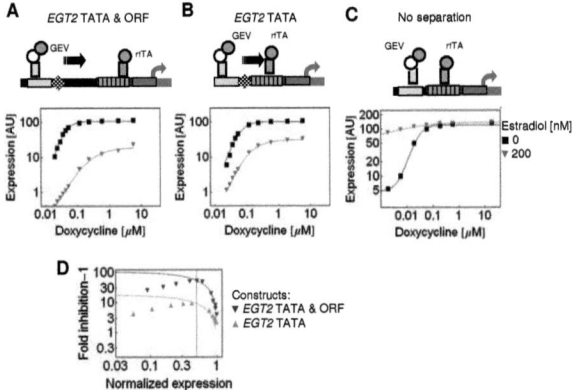

Figure S13. Initiation of upstream interference.
The downstream promoter is P$_{tetO7\text{-}GAL1TATA}$
(**A**) A truncated *EGT2* gene (*EGT2* -115 – 509), incorporating a TATA box, the abutting core promoter (-115 – 0) and a truncated ORF (1-509), separates the *GALUAS* and [*tetO*]$_7$ (YABH18.4).
(**B**) Only an 8 bp long TATA box separates the *GALUAS* and [*tetO*]$_7$. Expression is shown for constructs in the presence and absence of estradiol (YABH21.2).
(**C**) When GALUAS is directly fused to [tetO]$_7$. Estradiol activates gene expression (YABH20.1).
(**D**) Fold inhibition calculated for the data taken from (A) and (B) representing the YABH18.4 (*EGT2* -115 – 509), YABH21.2 (*EGT2* TATA) strains. The equilibrium competition model was used to fit Eq.1 (Box 1) to data points that had a normalized expression higher than 0.4 (see Materials and Methods). $K_D^A = 0.024$ and $f(R) = 18$, and 104 for YABH21.2 and -18.4, respectively.

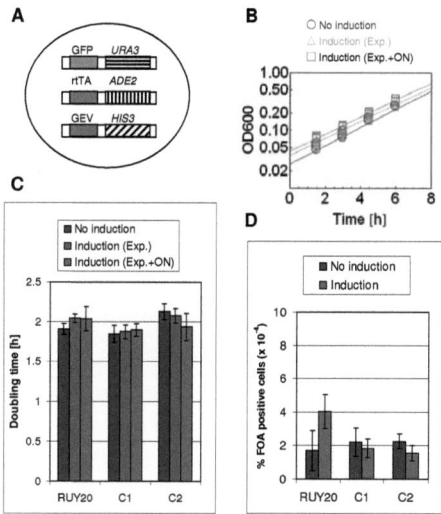

Figure S14. The effect of GEV and rtTA on cellular growth and on disintegration of chromosomally integrated gene constructs. When a transcriptional activator is expressed at high concentrations, it may bind to the components of the general transcriptional machinery, which leads to their depletion and to a decreased expression of cellular genes, a process termed squelching. Consequently, cellular growth may slow down. Furthermore, transcription can induce homologous recombination and the disintegration of the reporter constructs (Garcia-Rubio et al., 2003; Gonzalez-Barrera et al., 2002), which can affect the mean activity of the reporter gene in the cell culture. It is unclear how transcriptional interference affects recombination. To explore if squelching and transcription induced recombination affect our data, we tested how cellular growth and disintegration of the chromosomal constructs was influenced in the presence of GEV and rtTA.
(A) Genetic constructs with the linked marker genes are shown for the strain containing the $P_{GALUAS\text{-}TATA\text{-}tetO2}$ – GFP construct (RUY20). In this strain, the expression of GEV and rtTA were driven by promoters of moderate strength ($MRP7$ and $CLN3$). The control strains (YAntH61.1, 62.1) were obtained by integrating the $ADE2$ and $HIS3$ marker genes into the chromosome without the linked gene constructs, using empty pRS402 and pRS303 vectors.
(B) Triplicate measurements of cell densities are shown during the exponential phase of growth of RUY20. Cells from overnight cultures in selective medium (SD-Ade, -

Ura, -His) were inoculated into rich medium (YPD) at an $OD_{600} = 0.05$ to monitor the growth. Interference at maximal activation of gene expression was induced with 200 nM estradiol and 2μM doxycycline. The culture was induced either only during the exponential growth (Exp) or both during overnight growth (approximately 12 hours) and the exponential growth stages (ON + Exp). Curves were obtained by fitting the doubling (cell division) time using nonlinear regression.
(**C**). Error bars represent the 95% confidence interval for the fitted doubling rates (as shown in (B)). No significant differences in the doubling rates were observed neither between RUY20 and the control strains C1, 2 (YAntH60.1, 2), nor between the three different induction conditions for each of the strains.
(**D**) Disintegration of the URA3 marker was measured by counter-selecting Ura^+ cells on FOA (5-Fluoro-orotic acid) containing plates. Cells form overnight cultures in selective media (SD-Ade, -Ura, -His) were diluted into rich medium (YPD), grown for one day and again re-diluted in YPD and grown for another day. During this growth in non-selective conditions, approximately 20 cell divisions occurred before plating the cells (on plates containing 1 g / l FOA). The percentage of Ura^- colonies was calculated relative to the total number of colonies on YPD plates. Mean and standard deviation from three measurements are shown. A two-fold increase in recombination was observed when the YPD medium was supplemented with 200 nM estradiol and 2μM doxycycline. This increase affects less than 0.01% of the cells, thus it has a negligible contribution to the mean expression value of GFP.
Disintegration of the rtTA linked to *ADE2* can be evaluated by counting pink colonies on adenine poor plates. The percentage of these colonies was also less than 0.01% in all conditions (data not shown).

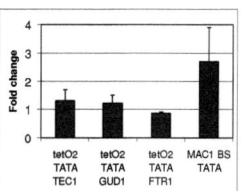

Figure S15. The effect of activator binding sites downstream of TATA boxes on gene expression. For a detailed exploration, we selected those genomic promoters (Table S1) that had predicted downstream binding sites for chemically inducible activators. Sequences downstream of the TATA boxes of the *TEC1*, *GUD1* and *FTR1* promoters contain predicted binding sites for the Ste12p, Gln3p and Mac1p transcriptional activators, which can be induced by α-factor, rapamycin and copper-depletion, respectively (Gross et al., 2000; Hagen et al., 1991; Kulkarni et al., 2006). The corresponding core promoters (TATA box and downstream sequence) were inserted downstream of two *tet* operators (YAntH64, 65; YAntH68, 71; YAntH66, 67). GFP expression was activated by rtTA (in the presence of 2μM doxycycline). Fold change in expression was measured 5 hours after exposing the cells to 1 μM α-factor, 0.2 μg/ml rapamycin and 100 μM BCS. BCS (bathocuproinedisulfonic acid) is a copper chelator, which depletes copper from the medium. Correction of expression using constructs with mutant binding sites was performed as for the constructs induced by α-factor shown in Figure 3D (see also Materials and Methods). Error bars represent standard deviations calculated from four experiments. Only the copper-depletion resulted in a small (~20%) but reproducible inhibition of gene expression induced by rtTA. When the Mac1 binding site was transferred upstream of a TATA box (MAC1BS-TATA, YAntH70), it elicited a 2.7 ± 1.2 fold activation of gene expression in the same conditions. Thus, the weak inhibition by MAC1 is consistent with the weak upstream activatory potential of this binding site. For comparison, rtTA elicits an approximately 30-fold activation of gene expression from an upstream site, and suppresses gene expression 3 - 4 times from a downstream site.

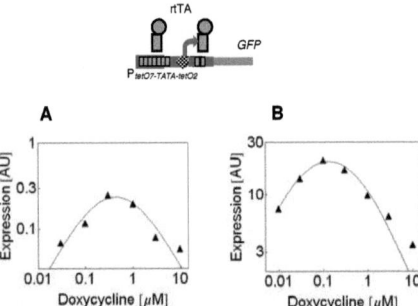

Figure S16. Bell-shaped response generated by the $P_{tetO7\text{-}TATA\text{-}tetO2}$ -GFP construct. (A) RUY53 (single copy of the construct in the genome). (B) RUY54 (~8 copies of the construct in the genome). Since expression driven by a single-copy construct is weak, lacZ was used to detect gene expression with higher sensitivity (Figure 3F).

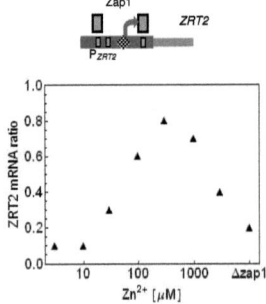

Figure S17. Bell-shaped response of the *ZRT2* gene to a gradient of zinc concentration. Data are taken from Figure 1 in [(Bird et al., 2004)]. High Zn^{2+} concentrations result in low Zap1 activity. Therefore, the conditions in the Δzap1 strain were approximated by that one in the wild-type strain exposed to 10000 μM Zn^{2+}.

168

Supplementary tables

Table S1. Conserved activator recognition sequences downstream of conserved TATA boxes.

Gene Name	SN[1]	C1[2]	TF[3]	TF Motif	Identity[4]	C2[5]	O[6]	BF[7]
GUD1	YDL238C	4	GLN3	GATAA	*******	4	F[a]	M
UTR2	YEL040W	4	NRG1	CCCT	*	3	F	S
UTR2	YEL040W	4	TEC1	.rCATTCy	*******	4	F	S
TEC1	YBR083W	4	DIG1	........tGAAAc.	* ************	4	F	S
TEC1	YBR083W	4	STE12	ATGAAAC	*******	4	F	S
PRB1	YEL060C	4	HAP5	CCAAT	*****	4	F	M
PRB1	YEL060C	4	HAP5	CCAAT	* *	3	F	M
PRB1	YEL060C	4	HAP5	CCAAT	**	3	R[b]	M
PRB1	YEL060C	4	MSN2	mAGGGG.	** ****	3	F	M
FTR1	YER145C	4	MAC1	GAGCAAA	* ***	3	F	S
SPS22	YCL048W	3	SUM1	...yGwCAswAA..	*********	3	R	S
PET8	YNL003C	3	AFT2	kgCACCc	*** ***	3	F	M
TIR3	YIL011W	3	HAP5	CCAAT	****	3	F	S
HSC82	YMR186W	4	INO4	gCATGTGAA	* ******	3	F	M
ICS2	YBR157C	4	NDD1	rAargGsAAA		3	R	M
SED1	YDR077W	4	MOT3	YAGGYA	**** *	3	R	S
SSA1	YAL005C	4	MSN2	MAGGGG	******	4	F	S
SSA1	YAL005C	4	MSN4	aAGGGG.	******	4	F	S
PIG2	YIL045W	3	CIN5	mtTAcrTAA	*********	4	F	S
ALK1	YGL021W	3	NDD1	rAargGsAAA	****** ***	3	R	S
CIN5	YOR028C	4	DAL80	GATAA	** **	3	F	M
MET13	YGL125W	2	MET32,4	mArcTGTGGC	**********	4	R	M

1 Gene Systematic Name
2 Number of Species in Which the TATA box is Conserved
3 Binding Transcription Factor
4 Downstream Motif Identity in Alignment Across the 4 Species
5 Number of Species in Which the Downstream Motif is Conserved
6 Downstream Motif Orientation
7 ChIP Binding Filter (M: Moderate Binding Filter, S: Stringent Binding Filter)
a Forward Orientation
b Reverse Complement Orientation

Conserved activator binding sites downstream of a TATA box were searched for, because transcriptional initiations sites are not well defined in *S. cervisiae,*. First, 585 genes were selected, which have a TATA consensus sequence in the core promoter regions (a 200 bp long sequence upstream of the ORF) and are conserved across four species (*S.cerevisiae, S.bayanus, S.paradoxus* and *S.mikatae*) (Basehoar et al., 2004). These genes were then subjected to TF binding site mapping, with the following criteria: motif conservation in four species, significance of binding based on ChIP-chip experiments and that TF motif is positioned downstream the TATA-box. The motif database contained data for 125 TFs (MacIsaac et al., 2006).

Table S2. Strain list

Strain	Description	Origin / Parent
RUY03.3	MAT α gal4Δ::kanMX, ura3::URA3 P$_{GALUAS\_TATA\_tetO2}$-YFP	pRU1, gal4Δ::kanMX → W303
RUY04, -05	MAT α ura3:: URA3_T$_{ADH1}$-P$_{tetO7\_CYC1TATA}$-YFP	pRU6 → W303
RUY08	MAT α his3::HIS3_P$_{MRP7}$_GEV (3-4 copies) gal4Δ::kanMX	pPR1, gal4Δ::kanMX → W303
RUY10	MAT α gal4Δ::kanMX, ura3::URA3 P$_{GALUAS\_TATA\_tetO2}$-YFP, his3::HIS3_ P$_{MRP7}$-GEV	pRU1, pPR1, gal4Δ::kanMX → W303
RUY20	MAT α gal4Δ::kanMX, ura3::URA3 P$_{GALUAS\_TATA\_tetO2}$-YFP, his3::HIS3_ P$_{MRP7}$-GEV, pRS402::P$_{CLN3}$-rtTA	pRS402::P$_{CLN3}$-rtTA → RUY10
RUY21	MAT A ade2::ADE2_ P$_{CLN3}$-rtTA	pRS402::P$_{CLN3}$-rtTA → W303
RUY22	MAT α ura3:: URA3_ T$_{ADH1}$-Ptet$_{O7\_TATA\_tetO2}$-lacZ	pRU2 → W303
RUY24.13	MAT α ura3:: URA3_T$_{ADH1}$-P$_{tetO7\_CYC1TATA}$-lacZ	pCS3 → W303
RUY28	MAT A ade2::ADE2_ P$_{CLN3}$-rtTA, gal4::kanMX	gal4Δ::kanMX → RUY21
RUY29	MAT α ura3:: URA3_ P$_{TATA\_tetO2}$-lacZ	pRU4 → W303
RUY53, -54	MAT A/α ura3:: URA3_ T$_{ADH1}$-Ptet$_{O7\_TATA\_tetO2}$-YFP-, ade2::ADE2_ P$_{SWI5}$-rtTA	ABY466b + RUY04, -05
RUY65	MAT A/α ura3:: URA3_ T$_{ADH1}$-Ptet$_{O7\_TATA\_tetO2}$-lacZ-, ade2::ADE2_ P$_{CLN3}$-rtTA	RUY21 + RUY22
RUY67.13	MAT A/α ura3:: URA3_T$_{ADH1}$-P$_{tetO7\_CYC1TATA}$-lacZ, ade2::ADE2_ P$_{CLN3}$-rtTA	RUY21 + RUY24.13
RUY69	MAT A/α ura3:: URA3_ P$_{TATA\_tetO2}$-lacZ, ade2::ADE2_ P$_{CLN3}$-rtTA	RUY21 + RUY29

Table S2. (continuation) Strain list

Strain	Description	Origin / Parent
YABH15.1	MAT α his3::HIS3_P_{MRP7}_GEV (3-4 copies) gal4Δ::kanMX, ade2::ADE2_ P_{CLN3}-rtTA	pRS402::P_{CLN3}-rtTA→ RUY08
YABH18.4	MAT α his3::HIS3_P_{MRP7}_GEV (3-4 copies) gal4Δ::kanMX, ade2::ADE2_ P_{CLN3}-rtTA, ura3:: URA3_ $P_{GALUAS\_EGT2TATA}$ –Egt2 trunc- $P_{tetO7-GAL1TATA}$ -YFP	pAB-BP10→ YABH15.1
YABH20.1	MAT α his3::HIS3_P_{MRP7}_GEV (3-4 copies) gal4Δ::kanMX, ade2::ADE2_ P_{CLN3}-rtTA, ura3:: URA3_ $P_{GALUAS\_EGT2TATA}$ –$P_{tetO7-GAL1TATA}$ -YFP	pAB-BP13→ YABH15.1
YABH21.2	MAT α his3::HIS3_P_{MRP7}_GEV (3-4 copies) gal4Δ::kanMX, ade2::ADE2_ P_{CLN3}-rtTA, ura3:: URA3_ $P_{GALUAS\_EGT2TATA}$ –$P_{tetO7-GAL1TATA}$ -YFP	pAB-BP12→ YABH15.1
YABH22.1	MAT A ade2::ADE2_ P_{CLN3}-rtTA, gal4::kanMX, his3::HIS3_P_{MRP7}-GEV	pPR1 → RUY28
YABH34.5	MAT A ade2::ADE2_ P_{CLN3}-rtTA, gal4::kanMX, his3::HIS3_P_{MRP7}-GEV, ura3:: URA3_ $P_{GALUAS\_EGT2TATA}$ –Egt2 trunc- $P_{tetO7-GAL1TATA}$ -YFP	pAB-BP10 → YABH22.1
YABH38.2	MAT A ade2::ADE2_ P_{CLN3}-rtTA, gal4::kanMX, his3::HIS3_P_{MRP7}-GEV, ura3:: URA3_ $P_{GALUAS\_EGT2TATA}$ –Egt2 trunc- $P_{tetO2-CYC1TATA}$ -YFP	pAB-BP46 → YABH22.1
YABH39.4	MAT A ade2::ADE2_ P_{CLN3}-rtTA, gal4::kanMX, his3::HIS3_P_{MRP7}-GEV, ura3:: URA3_ $P_{GALUAS\_EGT2TATA}$ –Egt2 trunc- $P_{tetO2-GAL1TATA}$ -YFP	pAB-BP47 → YABH22.1
YABH40.6	MAT A ade2::ADE2_ P_{CLN3}-rtTA, gal4::kanMX, his3::HIS3_P_{MRP7}-GEV, ura3:: URA3_ $P_{GALUAS\_EGT2TATA}$ –Egt2 trunc- $P_{tetO1-CYC1TATA}$ -YFP	pAB-BP45 → YABH22.1
YABH42.1	MAT A ade2::ADE2_ P_{CLN3}-rtTA,, gal4::kanMX, his3::HIS3_P_{MRP7}-GEV, ura3:: URA3_ $P_{GALUAS\_GAL1TATA::FUS1UAS}$-YFP	pAB-BP49 → YABH22.1
YABH43.2	MAT A ade2::ADE2_ P_{CLN3}-rtTA, gal4::kanMX, his3::HIS3_P_{MRP7}-GEV, ura3:: URA3_ $P_{GALUAS\_GAL1TATA::MutFUS1UAS}$-YFP	pAB-BP50 → YABH22.1
YABH50.31	MAT α his3::HIS3_PMRP7_GEV gal4Δ::kanMX ade2::ADE2_ PHTB2–rtTA-TCYC1 ura3:: URA3_PGALUAS–Egt2 trunc- PtetOinSIC1 –YFP (15-20 copies)	pAB-B54 → YABH46.1

172

Table S2. (continuation II) Strain list

Strain	Description	Origin / Parent
YAntH05.5	MAT α his3::HIS3_P_{MRP7}_GEV gal4Δ::kanMX ade2::ADE2_ P_{SWI5}–rtTA	pABT11 → RUY08
YAntH06.7	MAT α his3::HIS3_P_{MRP7}_GEV gal4Δ::kanMX ade2::ADE2_ P_{SWI5}–rtTA ura3:: URA3_P_{GALUAS}–Egt2 trunc- P_{EGT2} –YFP (7-copies)	pAB-BP55 → YAntH05.5
YAntH13.1	MAT α his3::HIS3_P_{MRP7}_GEV gal4Δ::kanMX ade2::ADE2_ P_{CLN3}–rtT-(NLS-AD)^{Swi5}	pAnt04 → RUY08
YAntH16.1	MAT α his3::HIS3_P_{MRP7}_GEV gal4Δ::kanMX ade2::ADE2_ P_{SWI5}–rtTA ura3:: URA3_P_{GALUAS}–Egt2 trunc- $P_{[tetO]6inEGT2}$ –YFP (4-copies)	pAB-BP57 → YAntH05.5
YAntH17.13	MAT α his3::HIS3_P_{MRP7}_GEV gal4Δ::kanMX ade2::ADE2_ P_{SWI5}–rtTA ura3:: URA3_P_{GALUAS}–Egt2 trunc- - $P_{[tetO]6inEGT2}$ –YFP (10-copies)	pAB-BP57 → YAntH13.1
YAntH41.2	MAT α his3::HIS3_P_{MRP7}_GEV gal4Δ::kanMX ade2::ADE2_P_{swi5}-rtTA ura3::URA3_GALUAS-Egt2trunc- $P_{[tetO]5*\_inEGT2}$-YFP	pAB-BP57M → YAntH05.5
YAntH42.2	MAT α his3::HIS3_P_{MRP7}_GEV gal4Δ::kanMX ade2::ADE2_P_{swi5}-rtTA, ura3::URA3_GALUAS-Egt2trunc—T_{ACT1}-$P_{[tetO]5*\_inEGT2}$-YFP	pAnt23 → YAntH05.5
YAntH43.1	MAT α his3::HIS3_P_{MRP7}_GEV gal4Δ::kanMX ade2::ADE2_P_{swi5}-rtTA, ura3::URA3_GALUAS-T_{ACT1} -$P_{[tetO]5*inEGT2}$-YFP	pAnt24 → YAntH05.5
YAntH44.6	MAT α his3::HIS3_P_{MRP7}_GEV gal4Δ::kanMX ade2::ADE2_P_{swi5}-rtTA ura3::URA3 GALUAS -P_{ADH1}-YFP	pAB-BP66 → YAntH05.5
YAntH61.1, 62.1	MAT α gal4Δ::kanMX, ura3::URA3 $P_{GALUAS\_TATA\_tetO2}$-YFP ade2::ADE2 his3::HIS3	pRS303, pRS402 → RUY03.3
YAntH64	MAT A ade2::ADE2_ P_{CLN3}-rtTA ura3::URA3 $P_{tetO2-TATA—TEC1ds}$ - GFP	pJK4 → RUY21
YAntH65	MAT A ade2::ADE2_ P_{CLN3}-rtTA ura3::URA3 $P_{tetO2-TATA—TEC1dsmut}$ - GFP	pJK14 → RUY21
YAntH66	MAT A ade2::ADE2_ P_{CLN3}-rtTA ura3::URA3 $P_{tetO2-TATA—FTR1ds}$ - GFP	pAnt47 → RUY21
YAntH67	MAT A ade2::ADE2_ P_{CLN3}-rtTA ura3::URA3	pAnt49 →

		RUY21
	$P_{tetO2-TATA-FTR1dsmut}$ - GFP	
YAntH68	MAT A ade2::ADE2_ P_{CLN3}-rtTA ura3::URA3 $P_{tetO2-TATA-GUD1ds}$ - GFP	pAnt47 → RUY21
YAntH70	MAT A ura3::URA3 $P_{tetO2-MAC1BS\ TATA}$ - GFP	pAnt50 → W303
YAntH71	MAT A ade2::ADE2_ P_{CLN3}-rtTA ura3::URA3 $P_{tetO2-TATA-GUD1ds}$ - GFP	pAnt48 → RUY21
YAntH96	MAT A ura3:: URA3_ $P_{GALUAS\_GAL1TATA::FUS1UAS}$-YFP	pAB-BP49 → W303 A
YAntH97	ura3:: URA3_ $P_{GALUAS\_GAL1TATA::MutFUS1UAS}$-YFP	pAB-BP50 → W303 A

Table S3. Plasmid list

Plasmid	Description	Construction
pRU1	pRS306::P$_{GALUAS\_TATA\_tetO2}$-YFP	{SalIxXhoI)-P$_{GAL10UAS\text{-}SphI\text{-}TATA\_tet[O2]2}$-BamHI-YFP-EcoRI-T$_{CYC1}$-NotI
pRU2	pRS306::T$_{ADH1}$-P$_{tetO7\text{-}TATA\text{-}tetO2}$-lacZ	XhoI-T$_{ADH1}$-P$_{tet[O2]7\text{-}SphI\text{-}TATA\_tet[O2]2}$-BamHI -lacZ-EcoRI-T$_{CYC1}$-NotI
pRU4	pRS306:: P$_{TATA\text{-}tetO2}$-lacZ	KpnI-Fig1p(180-897)-XhoI-T$_{GAL7}$-SphI-P$_{TATA\_tet[O2]2}$-BamHI lacZ-EcoRI- T$_{CYC1}$-NotI
pRU6	pRS306::T$_{ADH1}$-P$_{tetO7\text{-}TATA\text{-}tetO2}$-YFP	XhoI-T$_{ADH1}$-P$_{tet[O2]7\text{-}SphI\text{-}TATA\_tet[O2]2}$-BamHI -YFP-EcoRI-T$_{CYC1}$-NotI
pCS3	pRS306::T$_{ADH1}$-P$_{tetO7}$-lacZ	XhoI-T$_{ADH1}$-P$_{tet[O2]7\_CYC1TATA}$-BamHI- lacZ-EcoRI- T$_{CYC1}$-NotI
pPR1	pRS303::P$_{MRP7}$-GEV	P$_{MRP7}$-GEV Insert described in Gao & Pinkham, 2000: P$_{MRP7}$-GEV-T$_{Act1}$
pABT10	pRS402::P$_{CLN3}$-rtTA-T$_{CYC1}$	Becskei et al, 2005.
pABT11	pRS402::P$_{Swi5}$-rtTA-T$_{CYC1}$	Becskei et al, 2005.
pAB-BP10	pRS306:: P$_{GAL10UAS}$ - Egt2 trunc- P$_{tetO7\text{-}GAL1TATA}$ -YFP	ApaI-GAL10UAS-SphI-AvrII-Egt2 trunc-NheI – SalI- P$_{[tetO2]7\text{-}GAL1TATA}$-BamHI-YFP-EcoRI-T$_{CYC1}$-NotI
pAB-BP12	pRS306:: P$_{GAL10UAS}$ - P$_{tetO7\text{-}GAL1TATA}$ -YFP	ApaI-GAL10UAS-SphI-(AvrIIx NheI) – SalI- P$_{[tetO2]7\text{-}GAL1TATA}$-BamHI-YFP-EcoRI- T$_{CYC1}$-NotI
pAB-BP13	pRS306:: P$_{GAL10UAS}$ – EGT2TATA-P$_{tetO7\text{-}GAL1TATA}$ -YFP	ApaI-GAL10UAS-SphI-AvrII-EGT2 TATA-(SpeIxNheI) – SalI- P$_{[tetO2]7\text{-}GAL1TATA}$-BamHI-YFP-EcoRI- T$_{CYC1}$-NotI
pAB-BP45	pRS306:: P$_{GAL10UAS}$ - Egt2 trunc- P$_{tetO1\text{-}CYC1TATA}$ -YFP	ApaI-GAL10UAS-SphI-Egt2 trunc – SgsIx MluI P$_{[tetO2]1\text{-}SphI\text{-}CYC1TATA}$-(BamHIxBglII)-YFP-EcoRI- T$_{CYC1}$-NotI
pAB-BP46	pRS306:: P$_{GAL10UAS}$ - Egt2 trunc- P$_{tetO2\text{-}CYC1TATA}$ -YFP	ApaI-GAL10UAS-SphI-Egt2 trunc – SgsIx MluI P$_{[tetO2]2\text{-}SphI\text{-}CYC1TATA}$-(BamHIxBglII)-YFP-EcoRI- T$_{CYC1}$-NotI
pAB-BP47	pRS306:: P$_{GAL10UAS}$ - Egt2 trunc- P$_{tetO2\text{-}GAL1TATA}$ -YFP	ApaI-GAL10UAS-SphI-Egt2 trunc – SgsIx MluI P$_{[tetO2]2\text{-}SphI\text{-}GAL1TATA}$--(BamHIxBglII)-YFP-EcoRI- T$_{CYC1}$-NotI
pAB-BP49	pRS306::P$_{GALUAS\text{-}TATA\text{-}FUS1UAS}$-YFP	ApaI-P$_{GAL10UAS\_SphI\text{-}AvrII\text{-}}$

pAB-BP50	pRS306::P$_{GALUAS}$-TATA-MutFUS1UAS-YFP	GAL1TATA::FUS1UAS-BglII-YFP-EcoRI- T$_{CYC1}$-NotI ApaI-P$_{GAL10UAS\_SphI-AvrII-}$ GAL1TATA::MutFUS1UAS-BglII- YFP-EcoRI- T$_{CYC1}$-NotI
pAB-BP54	pRS306:: P$_{GALUAS\_EGT2TATA}$ –Egt2 trunc - P$_{tetOinSIC1}$ -YFP	SphI-AvrII- Egt2 trunc-NheI-SgsIxMluI- P$_{[tetO2]inSIC1}$-BglII -YFP-EcoRI- T$_{CYC1}$-NotI
pAB-BP55	pRS306:: P$_{GALUAS}$-Egt2 trunc- P$_{EGT2}$ -YFP	ApaI-GAL10UAS-SphI-Egt2 trunc – SgsIx MluI – P$_{EGT2}$-BglII-YFP-EcoRI- T$_{CYC1}$-NotI
pAB-BP57	pRS306:: PGALUAS_–Egt2 trunc - P$_{tetO6inEGT2}$ -YFP	KpnI-PGALUAS-SphI-AvrII-Egt2trunc-NheI-SgsI-P[tetO]6inEGT2 -BglII-YFP-NotI
pAB-BP57M	pRS306:: PGALUAS_ –Egt2 trunc - P$_{tetO5*inEGT2}$ -YFP	KpnI-PGALUAS-SphI-AvrII-Egt2trunc-NheI-SgsI-P[tetO]5*inEGT2 -BglII-YFP-NotI
pAB-BP66	pRS306: GALUAS -P$_{ADH1}$ -YFP	ApaI-PGAL10UAS_SphI-AvrII-PADH1-BglII- YFP-EcoRI-TCYC1-NotI
pAnt04	pRS306:: P$_{CLN3}$- rTetR-(NLS-AD)Swi5	PCLN3- BamHI-rTetR-(NLS-AD)Swi5-EcoRI- TCYC1-NotI
pAnt23	pRS306::GALUAS-Egt2trunc-TACT1-P$_{[tetO]5*inEGT2}$ -YFP	KpnI-PGALUAS-SphI-AvrII-Egt2trunc-NheI- TACT1-SgsI-P[tetO]5*inEGT2 -BglII-YFP-NotI
pAnt24	pRS306::GALUAS-TACT1-P$_{[tetO]5*inEGT2}$ -YFP	KpnI-PGALUAS-SphI-(AvrIIx NheI)- TACT1-SgsI-P[tetO]5*inEGT2 -BglII-YFP-NotI
pAnt47	pRS306::T$_{ADH1}$-P$_{tetO2}$-TATA—GUD1ds - GFP	XhoI-T$_{ADH1}$-P$_{tet[O2]2}$-SphI- TATA—GUD1ds - BamHI -GFP-EcoRI-T$_{CYC1}$-NotI
pAnt48	pRS306::T$_{ADH1}$-P$_{tetO2}$-TATA—GUD1dsmut - GFP	XhoI-T$_{ADH1}$-P$_{tet[O2]2}$-SphI- TATA—GUD1dsmut - BamHI -GFP-EcoRI-T$_{CYC1}$-NotI
pAnt49	pRS306::T$_{ADH1}$-P$_{tetO2}$-TATA—FTR1ds - GFP	XhoI-T$_{ADH1}$-P$_{tet[O2]2}$-SphI- TATA—FTR1ds - BamHI -GFP-EcoRI-T$_{CYC1}$-NotI
pAnt50	pRS306::T$_{ADH1}$-P$_{tetO2}$- MAC1BS TATA - GFP	XhoI-T$_{ADH1}$-P$_{tet[O2]2}$-SphI- MAC1BS TATA - BamHI -GFP-EcoRI-T$_{CYC1}$-NotI
pAnt51	pRS306::T$_{ADH1}$-P$_{tetO2}$-TATA—FTR1dsmut - GFP	XhoI-T$_{ADH1}$-P$_{tet[O2]2}$-SphI- TATA—FTR1dsmut - BamHI -GFP-EcoRI-T$_{CYC1}$-NotI
pJK4	pRS306::T$_{ADH1}$-P$_{tetO2}$-TATA—TEC1ds - GFP	XhoI-T$_{ADH1}$-P$_{tet[O2]2}$-SphI- TATA—TEC1ds - BamHI -GFP-EcoRI-

pJK14	pRS306::T$_{ADH1}$-P$_{tetO2}$-TATA—TEC1dsmut - GFP	T$_{CYC1}$-NotI XhoI-T$_{ADH1}$-P$_{tet[O2]2}$-SphI- TATA—TEC1dsmut - BamHI -GFP-EcoRI- T$_{CYC1}$-NotI

Table S4. Effect of antagonistic activators on the background gene expression.

Construct / Strain, Inducer [nM] ([Doxycyline] = D) ([Estradiol] = E) ([α-factor] = α)	Expression, Mean	Expression, Standard Deviation
P$_{tetO2\text{-}CYC1TATA}$, YABH38.2		
D= 0, E = 0	2.33	1.32
D= 0, E = 4	1.58	0.6
D= 0, E = 8	1.39	0.82
D= 0, E = 24	1.31	0.96
D= 0, E = 200	1.4	0.85
P$_{tetO2\text{-}GAL1TATA}$, YABH39.4		
D= 0, E = 0	0.85	0.56
D= 0, E = 4	0.49	0.32
D= 0, E = 8	0.36	0.49
D= 0, E = 24	0.25	0.64
D= 0, E = 200	0.25	0.62
P$_{GALUAS\text{-}TATA\text{-}tetO2}$, RUY20		
E= 0, D = 0	0.21	0.21
E= 0, D = 600	0.25	0.30
E= 0, D = 20,000	0.19	0.25
P$_{GAL UAS\text{-}TATA\text{-}FUSUAS}$, YAnt96, 97		
E= 0, α = 0	1.81	1.53
E= 0, α = 600	3.40	3.21

Sequences used for the gene constructs

T_{CYC1}, T_{GAL7}, T_{ADH1}, T_{ACT1} are transcriptional terminators of the respective genes.

EGT2trunc is a sequence that contains the EGT2 TATA box, an SpeI site, and a sequence downstream of the TATA box up to the 509 nucleotide of the Egt2 ORF: AvrII-ATATAAAAG-SpeI EGT2(-103 to 509)-NheI.

SWI5(4-1639) includes the activation and regulatory domain, while *SWI5*(1900-2125) incorporates a nuclear localization sequence (NLS), whose reversible phosphorylation is responsible for the import of Swi5p into the nucleus in the M and G1 stages of the cell cycle (Tebb et al., 1993).

GAL UAS

CCCCTCGACTTACTGCCAATTTTTCCTCTTCATAACCATAAAAGCTAGTAT
TGTAGAATCTTTATTGTTCGGAGCAGTGCGGCGCGAGGCACATCTGCGTTT
CAGGAACGCGACCGGTGAAGACGAGGACGCACGGAGGAGAGTCTTCCGT
CGGAGGGCTGTCGCCCGCTCGGCGGCTTCTAATCCGTACTTCAATAT

TATA-tet[O2]2

TATATAACCGGCAGATCAATTCCTCGA*TCCCTATCAGTGATAGAGA*GTCG
ACAAAGTCGAGTTTCTCGA*TCCCTATCAGTGATAGA*GAGTCGACAAAGTC
GAGTTTCTCGATCGAGACCACTGCAATACTCCATCCTTA

TATA box
TetR binding site [tetO2]
<u>Transcriptional initiator</u>

GAL1TATA-FUS1UAS

AGA**TATATAAA**TGCGATGA*TGAAACA*AACA*TGAAACG*TCAGTAATT*TGAA
ACA*A<u>ATTTT</u>CAGTTTGTATTACTTCTTATTCAAATGTCATAAAAGTATCAA
CAAAAAATTGTTAATATACCTCTATACTTTAACGTCAAGGAGAAAAAACT
AT

TATA box
Ste12p binding site (within the FUSUAS as described in (Hagen et al., 1991))

GAL1TATA-MutFUS1UAS

179

AGA**TATATAAA**TGCGATGA*TAGAAAC*AACA*TAGACAG*TCAGTAAT*TTAGA*
*AACA*ATTTTCAGTTTGTATTACTTCTTATTCAAATGTCATAAAAGTATCAA
CAAAAAATTGTTAATATACCTCTATACTTTAACGTCAAGGAGAAAAAACT
AT

TATA box
Mutated Ste12p binding site

$P_{[tetO]2inSIC1}$

*GATACGCGTTCCCTATCAGTGATAGAGA*CTGAAAAT*TTCCCTATCAGTGATA
GAGA*TTGTTCGCGTCGCAACTTTTGATAT**TTATAAAAA**GAGAGAAGGCAT
ATCCAAAGGATAAGGTATTGTCTTTGTAAATCAATCAACCAAACCTCTAC
GGAATTTTGACCCTTGAAGCAGGGACTATTACACGAAA

tetR binding site [tetO2]
TATA box

P_{EGT2}

ACTAATAT*AGCCTCTGCTGG*TTCCTTTAGTAACCTCTTTAGTAACCTCTTTA
GTGAAAT*GGTTCTTGCTGG*CTCTATTATTTTCCTAATT*CGGACGCGCTGG*C
TCCGCGCCGTGAGGAAAAA*CAGCAGGCTGA*CAAGGGACTAATTTACTGAC
*ACTTTCGGCTGA*CACTTCCGAAAAGGTACTCAAGCTTTTATGA*GTAAGATG
CTGG*TTTGATGCTAAGAATACGATTTAGTACTTCCTTTTTAATGTGGCTTGT
TTTTTTTTTATTCGTCCATAACAGTACAAA**TATAAAAG**CAGAATACATTCT
TGCCGACCTAAAACTTGTGTCCTTAGACTCAAACTTTCTGATGGTTTTCTTT
CAGTAGAATCAACATAAACAAGCATAATATAAAACACTACA

Swi5p / Ace2p binding site
TATA box

$P_{[tetO]5*inEGT2}$

ACTAATAT*TCCCTATCAGTGATAGAGA*AGTAACCTCTTTAGTAACCTCTTT
AGTGAA*TCCCTATCAGTGATAGAGA*TTATTTTCCTAATTCG*ACTCTATCATT
GATAGAGT*CGCCGTGAGGAAAA*T*CCCTATCAGTGATAGAGA*CTTCCGAAA
AGGTACTCAAGCTTTTA*TCCCTATCAGTGATAGAGA*TGCTAAGAATACGAT
TTAGTACTTCCTTTTTAATGTGGCTTGTTTTTTTTTTATTCGTCCATAACAGT
ACAAA**TATAAAAG**CAGAATACATTCTTGCCGACCTAAAACTTGTGTCCTT
AGACTCAAACTTTCTGATGGTTTTCTTTCAGTAGAATCAACATAAACAAGC
ATAATATAAAACACTACA

tetR binding site [tetO1, tetO2]
TATA box

P[tetO]6inEGT2

ACTAATAT*TCCCTATCAGTGATAGAGA*AGTAACCTCTTTAGTAACCTCTTT
AGTGAA*TCCCTATCAGTGATAGAGA*TTATTTTCCTAATTCG*ACTCTATCATT
GATAGAGT*CGCCGTGAGGAAAA*TCCCTATCAGTGATAGAGA*CTAATTTA*TC
CCTATCAGTGATAGAGA*CTTCCGAAAAGGTACTCAAGCTTTTA*TCCCTATC
AGTGATAGAGA*TGCTAAGAATACGATTTAGTACTTCCTTTTTAATGTGGCT
TGTTTTTTTTTTATTCGTCCATAACAGTACAAA**TATAAAAG**CAGAATACAT
TCTTGCCGACCTAAAACTTGTGTCCTTAGACTCAAACTTTCTGATGGTTTT
CTTTCAGTAGAATCAACATAAACAAGCATAATATAAAACACTACA

tetR binding site [tetO1, tetO2]
TATA box

                             TATA-*GUD1*ds (pAnt47)
SphI-
TTTCGGAT**TATAAAAG**GAGACA*AGATAAG*GATTTTAAAACAGGCCATTGG
CAAAAGAACAACATTTTATTGAATGAATATATGAATTGGGCTGTCCTTTGA
AGAGGATA-BamHI

TATA box
Gln3p Binding Site

                        TATA-*GUD1mut*ds (pAnt48)
SphI-
TTTCGGAT**TATAAAAG**GAGACA*GAAAATG*GATTTTAAAACAGGCCATTGG
CAAAAGAACAACATTTTATTGAATGAATATATGAATTGGGCTGTCCTTTGA
AGAGGATA-BamHI

TATA box
mutated Gln3p Binding Site

                          TATA-*FTR1*ds (pAnt49)
SphI-
GTTA**TATAAAA**TGCTCAATGT*GAGCAAA*AAAATACATAACACGACGGATT
GTACGTTGTCGAAATCTTCTCTCAGCAGGTCATCACACATATACTTCCCGC
C-BamHI

TATA box
Mac1p Binding Site

                    Mac1 binding site upstream TATA- (pAnt50)
SphI-
TCAATGT*GAGCAAA*AAAAAGCTTTTAGTTA**TATAAAA**TGCTACATAACAC

GACGGATTGTACGTTGTCGAAATCTTCTCTCAGCAGGTCATCACACATATA
CTTCCCGCC-BamHI

TATA box
Mac1p Binding Site

TATA-*FTR1mut*ds (pAnt51)
SphI-
GTT**ATATAAAA**TGCTCAATGT*AGGACAA*AAAATACATAACACGACGGATT
GTACGTTGTCGAAATCTTCTCTCAGCAGGTCATCACACATATACTTCCCGC
C-BamHI

TATA box
Mutated Mac1p Binding Site

TATA-TEC1ds (pJK4)
SphI-
TTT**TATAAAT**GTT*TATGAAAC*AGGAGATTCTTCGACTGGTGCCAGATTCCT
CCCATATCGACAAGAAGAATAATCCACCTATTTCAACAATTCTGATACCT
GTTTAACC-BamHI

TATA box
Ste12p Binding Site

TATA-TEC1ds (pJK14)
SphI-
TTT**TATAAAT**GTT*TAAAGCTT*AGGAGATTCTTCGACTGGTGCCAGATTCCT
CCCATATCGACAAGAAGAATAATCCACCTATTTCAACAATTCTGATACCT
GTTTAACC-BamHI

TATA box
Mutated Ste12p Binding Site

Supplementary references

Basehoar, A.D., Zanton, S.J. and Pugh, B.F. (2004) Identification and distinct regulation of yeast TATA box-containing genes. *Cell*, **116**, 699-709.

Becskei, A., Kaufmann, B.B. and van Oudenaarden, A. (2005) Contributions of low molecule number and chromosomal positioning to stochastic gene expression. *Nat Genet*, **37**, 937-944.

Bird, A.J., Blankman, E., Stillman, D.J., Eide, D.J. and Winge, D.R. (2004) The Zap1 transcriptional activator also acts as a repressor by binding downstream of the TATA box in ZRT2. *Embo J*, **23**, 1123-1132.

Evans-Galea, M.V., Blankman, E., Myszka, D.G., Bird, A.J., Eide, D.J. and Winge, D.R. (2003) Two of the five zinc fingers in the Zap1 transcription factor DNA binding domain dominate site-specific DNA binding. *Biochemistry*, **42**, 1053-1061.

Garcia-Rubio, M., Huertas, P., Gonzalez-Barrera, S. and Aguilera, A. (2003) Recombinogenic effects of DNA-damaging agents are synergistically increased by transcription in Saccharomyces cerevisiae. New insights into transcription-associated recombination. *Genetics*, **165**, 457-466.

Gonzalez-Barrera, S., Garcia-Rubio, M. and Aguilera, A. (2002) Transcription and double-strand breaks induce similar mitotic recombination events in Saccharomyces cerevisiae. *Genetics*, **162**, 603-614.

Gross, C., Kelleher, M., Iyer, V.R., Brown, P.O. and Winge, D.R. (2000) Identification of the copper regulon in Saccharomyces cerevisiae by DNA microarrays. *J Biol Chem*, **275**, 32310-32316.

Hagen, D.C., McCaffrey, G. and Sprague, G.F., Jr. (1991) Pheromone response elements are necessary and sufficient for basal and pheromone-induced transcription of the FUS1 gene of Saccharomyces cerevisiae. *Mol Cell Biol*, **11**, 2952-2961.

Kenakin, T. (2004) Principles: receptor theory in pharmacology. *Trends Pharmacol Sci*, **25**, 186-192.

Kulkarni, A., Buford, T.D., Rai, R. and Cooper, T.G. (2006) Differing responses of Gat1 and Gln3 phosphorylation and localization to rapamycin and methionine sulfoximine treatment in Saccharomyces cerevisiae. *FEMS Yeast Res*, **6**, 218-229.

MacIsaac, K.D., Wang, T., Gordon, D.B., Gifford, D.K., Stormo, G.D. and Fraenkel, E. (2006) An improved map of conserved regulatory sites for Saccharomyces cerevisiae. *BMC Bioinformatics*, **7**, 113.

Schaufler, L.E. and Klevit, R.E. (2003) Mechanism of DNA binding by the ADR1 zinc finger transcription factor as determined by SPR. *J Mol Biol*, **329**, 931-939.

Tebb, G., Moll, T., Dowzer, C. and Nasmyth, K. (1993) SWI5 instability may be necessary but is not sufficient for asymmetric HO expression in yeast. *Genes Dev*, **7**, 517-528.

Wolner, B.S. and Gralla, J.D. (2001) TATA-flanking sequences influence the rate and stability of TATA-binding protein and TFIIB binding. *J Biol Chem*, **276**, 6260-6266.

Part III
The Role of Promoters within Small Regulatory Networks

1 Introduction

> *In this last part, a brief description of a currently ongoing project is given. The participation on my side is restricted to an initial phase of the project that has been developed further by other members of the group. The current project is thereby subject to modifications and further interpretation based on more recent and future findings. I anyway decided to integrate the work I contributes to in this short part summarizing ideas and goals of the project at an initial stage of its development.*

In third and last part of this work, the role of promoters in small gene regulatory networks has been investigated. We chose the galactose network of *Saccharomyces cerevisiae* (GAL network) and in particular we focused on the *GAL1* and *GAL3* genes. These genes are coregulated by the activator Gal4p and related by an evolutionary process involving the optimization of their functions. An ancestral bifunctional gene (*GAL1/3*) still present in *Kluyveromyces lactis*, a yeast species that didn't undergo whole genome duplication, gave rise to the two actual forms of *GAL1* and *GAL3* genes present in modern *Saccharomyces cerevisiae*. *GAL3* function of galactose sensing and *GAL1* galactokinase activity originated from a process of gene duplication followed by subfunctionalization of the ancestral sequence of *GAL1/3*, a gene coding for a protein executing both functions. After duplication, the previously disfavored binding site configuration containing four Gal4p binding sites evolved into two specialized gene versions containing four and one Gal4p binding sites in *GAL1* and *GAL3* genes, respectively (see Fig.56). These two genes ultimately became one of the most tightly regulated genes in the genome [7]. The different properties of *GAL1* and *GAL3* regulation have been studied in this part of the work in the context of the GAL network.

The GAL network displays bistable mode of response to galactose induction due to feedback loops. The parameters characterizing retroactive regulation are determining for the ON/OFF state transition of the network [114]. The results obtained in the previous parts of this work on the effect of variable number of activator binding sites in promoters have been related to the context of the GAL network as well as the role of stochastic network state transitions due to basal expression. In order to imitate GAL network behavior with respect to parameters such a cooperativity in activation and basal expression level, we engineered a synthetic equivalent network where we replaced native binding sites and activator with the synthetic rtTA–*tet*-operators binding partners. The strength of the positive feedback loop in the synthetic network could thereby be adjusted externally and properties of different promoter structures studied in relation to the efficiency in switching the state of the network.

Finally, inspired by the information obtained with help of the synthetic network, we modified the promoter of the native *GAL3* gene and replaced the endogenous copy present in the yeast genome. This allowed to experimentally evaluate the adaptation of cells to changing nutrient condition in relation to different gene constructs responsible for nutrient sensing.

2 Method

2.1 Experimental Methods

2.1.1 Promoters

An artificial *GAL1* promoter containing three binding sites has been obtained by substituting the three stronger distal binding sites for Gal4p by *tet*-operators (named *tetO3inGAL1*). Another version of the *GAL1* promoter has been obtained by substituting only the proximal binding site (named *tetO1inGAL1*) and representing an intermediate connecting the two stages of the evolutionary process that gave rise to the actual *GAL1* and *GAL3* promoter sequences. Finally a construct imitating *GAL3* has been obtained by substitution of the unique binding site for Gal4p by a *tet*-operator (named *tetO1inGAL3*). All of these modified promoters drove the expression of rtTA representing a positive feedback loop whose strength could be adjusted by doxycycline concentration (see Fig.56). The constructs have been transformed in a $\Delta GAL4$ BY4741 strain containing a reporter gene activated by two *tet*-operators and induced by doxycycline in glucose free conditions (2% raffinose and 0.5% galactose) for 24 hrs. The samples were analyzed by flow cytometry and the characteristic bimodal cell population evaluated by gating the fraction of cells being in activated (ON) state with respect to the whole cell population.

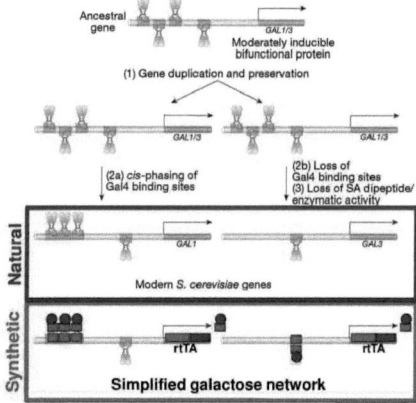

Figure 56: Scheme of the process of gene duplication and subsequent subfunctionalization underlying evolution of *GAL1* and *GAL3* from an ancestral bifunctional gene (*GAL1/3*) [7]. The part within the red frame represents the synthetic network used in our work to mimic the behavior of the GAL network. *tetO3inGAL1* consists of the *GAL1* promoter architecture with the three strong distal binding sites for Gal4p substituted by *tet*-operators. *tetO1inGAL1* has been obtained by substituting only the proximal binding site. In *tetO1inGAL3* the unique binding site for Gal4p has been substituted. The figure is taken and modified from [7].

2.1.2 Basal Expression Measurement

The basal expression value corresponding to the different promoters and supposed to be relevant for the network switching efficiency was measured by real-time pcr under non inducing conditions.

2.1.3 Adaptation

Adaptation experiments were performed by inducing with 0.5% galactose concentration exponentially growing cells containing the different promoter versions driving expression of the wild type $GAL3$ gene. Cell growth was monitored over 11 hrs by OD_{600} and compared to non induced cells.

2.2 Stochastic Simulations

The parameters obtained experimentally were used to run stochastic simulations (Gillespie algorithm, direct method [115]) of the model shown in Fig.57. The parameters that were not measured in our work have been taken from the literature [116]. The model described in Fig.57 accounts for transcription, translation and nuclear transport of rtTA. Nuclear rtTA induces its own transcription in a graded or switch-like manner according to the Hill function $\frac{[rtTA]^{n_H}}{[rtTA]^{n_H}+K^{n_H}}$ where K is considered to be a dissociation constant. An additional reaction module accounts for the basal expression level and its activity generates bursts of mRNA with different frequencies.

Method

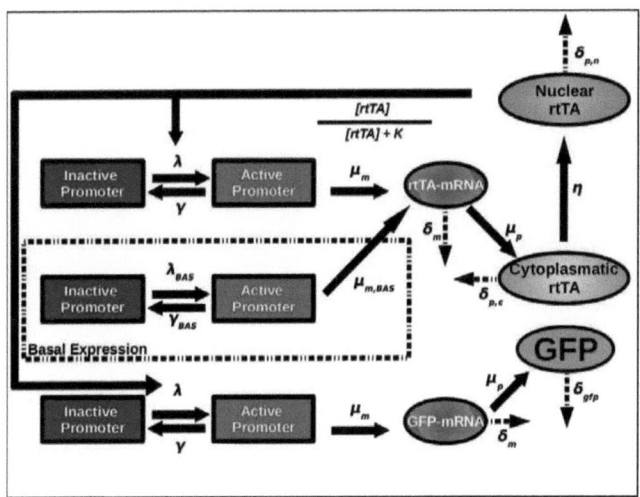

Figure 57: Reaction scheme of the stochastic simulation used to reproduce ON/OFF transitions of the bistable synthetic GAL network. K was varied according to variable induction level. No binding cooperativity of rtTA was assumed for $tetO1inGAL1$ and $tetO1inGAL3$ ($n_H = 1$) while $tetO3inGAL1$ was assumed to be positively cooperative ($n_H = 2$) (according to the results obtained in 'Full-Synthetic Systems' ('Part I')). $\lambda_{BAS}, \gamma_{BAS}, \mu_{m,BAS}$ were combined according to the formula: $\langle mRNA \rangle = \frac{\lambda_{BAS}}{\lambda_{BAS}+\gamma_{BAS}} \cdot \frac{\mu_{m,BAS}}{\delta_m}$ in order to obtain the mean steady state level of mRNA corresponding to the experimentally measured one. Similarly, λ, γ, μ were combined according to the formula: $\langle mRNA \rangle = \frac{\lambda}{\lambda+\gamma} \cdot \frac{\mu_m}{\delta_m}$ in order to obtain a mean steady state level of mRNA representing maximal expression measured experimentally under high induction conditions.

Results and Discussion

3 Results and Discussion

Cells were grown over night under non inducing conditions. When induced with doxycycline, $tetO1inGAL3$ allowed a faster ON state transition than $tetO3inGAL1$ (see Fig.58, left panel). Running stochastic simulations of the model shown in Fig.57 with high basal expression and assuming uncooperative binding of rtTA, the fast transition pattern of $tetO1inGAL3$ could be reproduced. In contrast, $tetO3inGAL1$ was simulated accounting for a roughly 10-fold lower basal expression value and high cooperativity ($n_H = 2$) resulting in higher memory of the previously applied growing conditions. Different frequencies in the simulation modules accounting for basal and promoter activation could influence to a certain extent the transition curve of $tetO3inGAL1$ and resulted similar to the curve obtained experimentally corresponding to $tetO1inGAL1$.

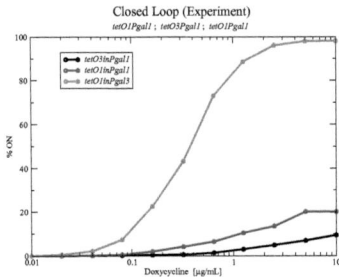 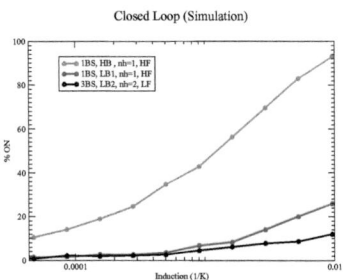

Figure 58: Left panel: Experimentally measured state transition curves of the closed feedback loop regulated by $tetO3inGAL1$, $tetO1inGAL1$, $tetO1inGAL3$ driving the expression of rtTA. The y-axis represents the percentage of the entire cell population expressing high levels of a reporter gene induced by rtTA. Right panel: Stochastic model representing the closed feedback loop corresponding to the synthetic GAL network. $\lambda_{BAS}, \gamma_{BAS}, \mu_{m,BAS}$ were combined according to the formula: $\langle mRNA \rangle = \frac{\lambda_{BAS}}{\lambda_{BAS}+\gamma_{BAS}} \cdot \frac{\mu_{m,BAS}}{\delta_m}$ in order to obtain the steady state level of mRNA corresponding to the experimentally measured one. The basal expression steady state value was set to 0.0135 and 0.126 for low and high basal expression, respectively. The promoters $tetO3inGAL1$ and $tetO1inGAL1$ had low basal expression level while $tetO1inGAL3$ high basal expression (HB, $\lambda_{BAS} = 0.034, \gamma_{BAS} = 54, \mu_{m,BAS} = 20$). Two different frequencies of bursting have been used for the low basal expression level: LB1 (low basal 1, $\lambda_{BAS} = 0.034, \gamma_{BAS} = 5, \mu_{m,BAS} = 0.2$) and LB2 (low basal 2, $\lambda_{BAS} = 0.00034, \gamma_{BAS} = 5, \mu_{m,BAS} = 20$). λ, γ, μ were combined according to the formula: $\langle mRNA \rangle = \frac{\lambda}{\lambda+\gamma} \cdot \frac{\mu_m}{\delta_m} = 20$ in order to obtain the steady state level of mRNA with respect to the basal expression with two different promoter activation frequencies named LF (low frequency, $\lambda = 0.75, \gamma = 6.75, \mu = 20$) and HF (high frequency, $\lambda = 6, \gamma = 54, \mu = 20$). The reporter promoter was parametrized with $\lambda = 5, \gamma = 20, \mu = 5$. The rest of the parameter values was taken from [116].

Results and Discussion

The promoter structures studied with the synthetic network context suggested a rapid adaptation to changing nutrient conditions supported by the *GAL3* promoter. Corresponding constructs were thereby made incorporating native Gal4p binding sites in the promoters of *GAL1* or *GAL3* and driving expression of *GAL3* ORF in the natural GAL network context. In addition, a version of the *GAL1* promoter containing only the distal Gal4p binding site has been compared to the wild type *GAL1* and *GAL3* promoter in order to account for a potentially intermediate step during the evolutionary process of gene duplication and subfunctionalization. The three different constructs were substituted to the endogenous *GAL3* promoter and regulated expression of galactose sensing protein of the GAL network (Gal3p). The strains containing the different constructs were induced by galactose (0.5%) and their growth monitored over 11 hrs. Cells containing the wild type version of the *GAL3* promoter could grow most efficiently, cells containing *GAL1* promoter the least and an intermediate response was observed by the *GAL1* promoter containing a single Gal4p binding site (see Fig.59).

The results showed that the low level of memory displayed by *GAL3* promoter allowed the network system to adapt more rapidly to changing nutrient conditions with respect to *GAL1* promoter. Moreover, *GAL1* promoter containing a single activator binding site displayed an intermediate response with respect to the two other constructs.

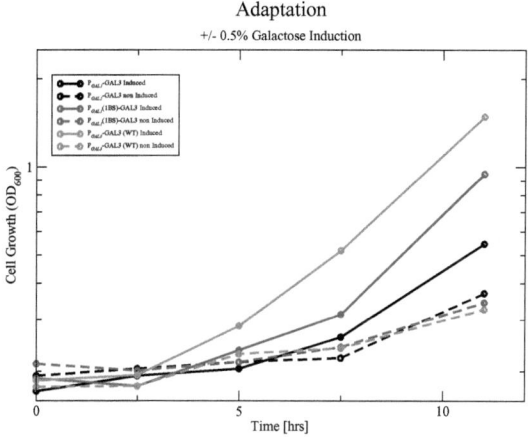

Figure 59: Adaptation experiment. The constructs analyzed in the context of the synthetic network have been used to replace the promoter of the endogenous *GAL3* gene and cell growth was measured in a time course of 11 hrs after 0.5% galactose induction (time point 0).

4 Conclusions

The different promoter versions derived from the yeast's *GAL1* and *GAL3* genes showed significantly different behavior to inducing conditions and pointed out the determining role played by cooperativity on one hand, and by the basal expression level inducing stochastic state transitions of the bistable system on the other hand. The use of a synthetic network to imitate the natural one allowed to identify key parameters determining state transition efficiency. The information obtained in this way could be transferred to the context of the GAL natural network and parameters characterizing the studied promoters could be linked to cellular adaptability to changing conditions.

The *GAL3* promoter resulted to be best optimized to rapid state transitions as a consequence its high basal expression level, while lower basal expression and high cooperativity increased memory of *GAL1* promoter thereby reducing its adaptation capacity to changing nutrient conditions.

Overall Conclusions

The use of synthetic gene circuits introduced into native cellular environment allowed to access information about different studied systems. Applying mathematical modeling approaches we could interpret our experimental findings. This approach has been applied to different levels of organization from single transcriptional units, to interacting transcriptional processes and finally to interconnected genes in the context of small regulatory networks. The work presented in this thesis was focused on the role of cooperativity in the process of gene regulation. Even if genes are thought to be expressed in all-or-none fashion as a consequence of their network connectivity, we observed different degrees of cooperativity and graded response. Our results show that role of cooperativity in gene regulation is relevant at the three different studied levels of complexity.

Cooperative binding of regulators to promoter sequences is mediated by components of the transcriptional machinery. The case of $GAL1$ shows that a switch-like response is achieved at low induction, while at higher activity the promoter displays a graded response pattern thereby conserving the ability to apply a proportionate response to higher induction levels. Multiple activator binding sites contributed in a more-than-additive increase of expression over a broad range of induction compared to the promoter activated by a single activator binding site. This supports the idea that eukaryotic activators can cooperate not by directly interacting but by simultaneously touching some components of the transcriptional machinery.

Further, we studied interacting transcriptional processes in the context of transcriptional interference. The process of transcriptional interference is commonly thought to have a suppressive effect on gene expression. Our results show that cooperativity between transcriptional processes could reverse the suppressive effect typically associated to interference and result in activation.

In the highest level of organization studied in this work, cooperativity together with the role of promoter basal expression were involved in increasing memory of bistable synthetic regulatory networks. High cooperativity and low basal expression dictating $GAL1$ regulation had the effect to increase its memory to previously applied conditions. In contrast, the regulation of the GAL network sensor, Gal3p, showed a reduced memory thus improving its adaptability to changing environmental conditions.

Acknowledgements

General Acknowledgements

First and foremost I am grateful to those who helped me complete this work. To Prof. Dr. Attila Becskei, for putting his trust in me and allowing me to work in his group at the University of Zurich and to all the people in the group: Janos Kelemen and Chetak Shetty for teaching all kind of theoretical and experimental skills and tricks in a highly motivating atmosphere, for reading the manuscript providing help and useful suggestions. Simone Scherrer for the Indian spirit and the great patience during teaching me all experimental aspects of the wet lab work. Prof. Dr. Antonio Baici for the excellent scripts of enzyme kinetics and ligand binding, and also for giving the best lectures in the 8 years of university in Zurich as well for the support and useful discussions. Marie Mi Bonde Hansen, Chieh Hsu, Aurelio de los Reyes, Vincent Jaquet for reading and correcting the manuscript as well as Firouzeh Farahmand, Mümün Gencoglu, Shweta Gujar, Li Sun for motivation and support through the fantastic atmosphere created in the lab like in the coffee breaks and TGIF.

Acknowledgements for Part I

Besides the supervision of Attila Becskei to the work, Shweta Gujar, Mümün Gencoglu and Li Sun contributed to 'Part I'. Their findings have however not been included in this thesis but rather represent a continuation of the (ongoing) work presented here. Ideas and useful discussions with them have largely been inspiring the development of 'Part I'. Very crucial and important discussions especially involving Janos Kelemen and Chetak Shetty were determining for the accomplishment of the work presented in this part.

Acknowledgements for Part II

The publication of the article attached in the second part could only be possible with the collaboration of several people. First of all Attila Becskei who started this project before I joined the group together with Rosemarie Ungricht and Chetak Shetty. Attila Becskei conceived the modeling that could explain most of our experimental findings and wrote the article. Janos Kelemen performed the bioinformatics search of conserved activator recognition sequences downstream of TATA boxes. Rosemarie Ungricht and Chetak Shetty performed experiments together with Attila Becskei for the downstream interference and bell-shaped response. Prasuna Ratna, Simone Scherrer performed additional experiments and Beatrice Blattmann technical help.

Acknowledgements for Part III

The third part of the work presented in this thesis largely represents ongoing and incomplete work. Hence, I have to acknowledge Chieh Hsu, Vincent Jaquet and Attila Becskei for the continuation of the project. The brief part of the 'Part III' project presented in this thesis has been presented according to the current knowledge and experimental situation of the time point where I was still participating to it (roughly until beginning of 2011). During this period I have to acknowledge Simone Scherrer, Prasuna Ratna for experiments and Firouzeh Farahmand for technical help.

Glossary

Transcription Unit: DNA sequence that contains the open reading frame (ORF) that will be translated into the protein (the coding sequence) and the regulatory sequences that direct and regulate the synthesis of that protein.

Chemical Equilibrium: State in which the chemical activities or concentrations of the reactants and products have no net change over time, all chemical potential gradients are zero (detailed balance holds for each of the reactions, the rate at which any process proceeds in the 'forward' direction is exactly balanced by the rate of that process in the 'reverse' direction). At dynamic equilibrium there can be no net flux of the system components. Moreover, the equilibrium state of a chemical reaction is unique (e.g. no oscillations possible in equilibrium chemistry). A direct consequence of dynamical systems is that eventually, every isolated chemical system should reach its equilibrium (e.g. death). Equilibrium is a special case of a steady state.

Steady State: Steady state refers to a system in which some components are kept constant at a stationary state but not at the values expected for the equilibrium state (detailed balance doesn't hold). A system in steady state is time invariant. In stochastic systems, the probabilities that various different states will be repeated will remain constant. While a dynamic equilibrium occurs when two or more reversible processes occur at the same rate, and such a system can be said to be in steady state, a system that is in steady state may not necessarily be in a state of dynamic equilibrium, because some of the processes involved are not reversible. For example: The flow of fluid through a tube, or electricity through a network, could be in a steady state because there is a constant flow of fluid, or electricity. Conversely, a tank which is being drained or filled with fluid would be an example of a system in transient state, because the volume of fluid contained in it changes with time.

Chemical Kinetics (or Reaction Kinetics): The study of rates of chemical processes and the different experimental conditions that can influence the speed of a chemical reaction yielding information about the reaction's mechanism and transition states, as well as the construction of mathematical models that can describe the characteristics of a chemical reaction.

Law of Mass Action: States that the speed of a chemical reaction is proportional to the quantity of the reacting substances and its propensity to occur.

Rapid-Equilibrium Assumption: Reaction rates like complex formation in reaction kinetics are assumed to be orders of magnitude faster with respect to other reactions (e.g. chemical conversion of substrate to products) so that chemical equilibrium can be assumed between fast reactions and rate-limiting steps are determined by slow reaction.

Steady-State Assumption: The intermediate species in chemical reactions are considered to be at steady-state and thereby they do not change in time. Whereby one considers the dynamics of the enzyme-substrate complex to be so fast that its concentration can be treated as if it were in steady state. This is equivalent to setting the time derivative of the enzyme-substrate concentration to zero and reducing the differential equation to an algebraic one [76].

Core Promoter: The minimal portion of the promoter required to properly initiate transcription. Contains the TATA-box at about 25 bp upstream of the start site to which general transcription factor binding sites bind. It also contains a pyrimidine-rich initiator element (located at the start site).

Promoter-Proximal Element: Any regulatory sequence in eukaryotic DNA that is located approximately 50-200 bp upstream of the start site and binds specific transcription factors.

Promoter-Distal Element: Sequence further upstream of the promoter that may contain additional specific transcription factor binding sites, often with a weaker influence than the proximal promoter and constitute another group of DNA targets for factors modulating RNA polymerase II activity.

Enhancers: Enhancers are long-range activators of gene transcription in higher eukaryotes. Enhancer and promoter elements can anyway overlap both physically and functionally. Enhancers positively regulate promoters in a distance and orientation independent manner. In this process they are at close three dimensional proximity despite they can be separated by thousands of base pairs. Their elements are composed of a modular arrangement of short sequence motifs with specific function in transcription that contain binding sites for nuclear activators [117, 118].

Transcriptional Regulators: Proteins that upon binding to DNA regulatory regions are capable of modulating the process of transcription and consequently influence gene expression. The generic term regulator refers to activator and repressors of transcription.

Binding Event: Microscopic binding reaction in a sequential process of n binding steps.

Receptor Configuration: Molecular state of a receptor bound by a certain number of ligands at equilibrium. Under the conditions assumed for the binding reaction, the configuration of a receptor refers to the population of molecules bound by the same number of ligands.

Low/High Order Binding Event: In the present work, upon multiple activator/ligand binding to a receptor, the concept of binding order is introduced accordingly to the exponents of Adair's equations where order of binding includes binding events that form receptor configurations containing certain number of bound ligands. Typically in the present work, when mentioning low binding order refers to binding events that form receptor configuration containing 2/3 activators while high binding refers to more activators.

Consensus Sequence: DNA sequence specifically recognized by transcriptional regulators. These sequences usually contain portion that are more important for protein-DNA binding (representing some key contacts between the regulator and the DNA) while some other portion have a more negligible effect and can thereby be subject to mutations resulting in slightly modulating activator affinity. Typically, such regulators are homodimers and consensus sequences are palindromic.

Synergic System: Synergy and cooperativity are generally used as synonyms and sometime this denomination is misleading. Actually, in this work the difference is made between *binding* cooperativity, which refers to the mathematical definition developed in the field of enzyme kinetics and molecular binding. In contrast, synergy refers to a system displaying response where the combination of different parts has an effect that is larger than the sum of effects displayed by single parts separately. Here however, no mathematical formalism is detailed.

Input/Output: In the context of gene regulation, input corresponds to induction or stimulus, that invokes a response or output of the involved system.

Saturation Functions: Mathematical expression that describes the degree of activation of a system as a function of its inputs. In engineering these functions are also called transfer functions and they relate output to input in a system. In chemical kinetics they can describe enzyme activity, occupancy of a receptor molecule by its ligands or level of gene activation as a function of a component of the reaction.

Gene Activation Profile: Indicates the same concept represented by 'Saturation Function' but in the specific context of gene regulation. It indicates the activity level of a gene system over a certain range of induction.

Dissipative systems: Systems that are not energy conserving. Perturbations decay until they get completely lost. Dissipative systems needs permanent energy (e.g. food) supply. As long as they get it, they are out of thermodynamic equilibrium. Only after the supply stops, they are subject to equilibrium thermodynamics (e.g. decay, second law). Dissipative systems are not reversible. If a system is on an attractor, it's not possible to reconstruct its history (if the state can olny be measured with finite precision).

Parabolic Partial Differential Equation: Type of second-order partial differential equation (PDE), describing a wide family of problems in science (e.g. heat diffusion). These problems, also known as evolution problems, describe physical or mathematical systems with a time variable, and which behave essentially like heat diffusing through a solid.

Hyperbolic Partial Differential Equation: Partial differential equation (PDE) that, roughly speaking, has a well-posed initial value problem for the first $n-1$ derivatives.

References

[1] Mukherji S, van Oudenaarden A. Synthetic biology: understanding biological design from synthetic circuits. *Nat Rev Genet*, 10(12):859–71, 2009.

[2] Poritz MA, Malmstrom S, Kim MK, Rossmeissl PJ, Kamb A. Graded mode of transcriptional induction in yeast pheromone signalling revealed by single-cell analysis. *Yeast*, 18(14):1331–8, 2001.

[3] Takahashi S, Pryciak PM. Membrane localization of scaffold proteins promotes graded signaling in the yeast map kinase cascade. *Curr Biol*, 18(16):1184–91, 2008.

[4] Fiering S, Whitelaw E, Martin DI. To be or not to be active: the stochastic nature of enhancer action. *Bioessays*, 22(4):381–7, 2000.

[5] Biggar SR, Crabtree GR. Cell signaling can direct either binary or graded transcriptional responses. *EMBO J*, 20(12):3167–76, 2001.

[6] Anders A, Lilie H, Franke K, Kapp L, Stelling J, Gilles ED, Breunig KD. The galactose switch in kluyveromyces lactis depends on nuclear competition between gal4 and gal1 for gal80 binding.

[7] Hittinger CT, Carroll SB. Gene duplication and the adaptive evolution of a classic genetic switch. *Nature*, 449(7163):677–81, 2007.

[8] Ellis T, Wang X, Collins JJ. Diversity-based, model-guided construction of synthetic gene networks with predicted functions. *Nat Biotechnol*, 27(5):465–71, 2009.

[9] Griggs DW, Johnston M. Regulated expression of the gal4 activator gene in yeast provides a sensitive genetic switch for glucose repression. *Proc Natl Acad Sci U S A*, 88(19):8597–601, 1991.

[10] Giorgetti L, Siggers T, Tiana G, Caprara G, Notarbartolo S, Corona T, Pasparakis M, Milani P, Bulyk ML, Natoli G. Noncooperative interactions between transcription factors and clustered dna binding sites enable graded transcriptional responses to environmental inputs. *Mol Cell*, 37(3):418–28, 2010.

[11] Lahav G, Rosenfeld N, Sigal A, Geva-Zatorsky N, Levine AJ, Elowitz MB, Alon U. Dynamics of the p53-mdm2 feedback loop in individual cells. *Nat Genet*, 36(2):147–50, 2004.

[12] Ferrell JE Jr; Machleder EM. The biochemical basis of an all-or-none cell fate switch in xenopus oocytes. *Science*, 280(5365):895–8, 1998.

[13] A Johnson, B J Meyer, and M Ptashne. Mechanism of action of the cro protein of bacteriophage lambda. *Proc Natl Acad Sci U S A*, 75(4):17831787, 1978.

[14] Darling PJ, Holt JM, Ackers GK. Coupled energetics of lambda cro repressor self-assembly and site-specific dna operator binding ii: cooperative interactions of cro dimers. *J Mol Biol*, 302(3):625–38, 2000.

[15] Ackers GK, Johnson AD, Shea MA. Quantitative model for gene regulation by lambda phage repressor. *Proc Natl Acad Sci U S A*, 79(4):1129–33, 1982.

[16] Shea MA, Ackers GK. The or control system of bacteriophage lambda. a physical-chemical model for gene regulation. *J Mol Biol*, 181(2):211–30, 1985.

[17] Becskei A, Séraphin B, Serrano L. Positive feedback in eukaryotic gene networks: cell differentiation by graded to binary response conversion. *EMBO J*, 20(10):2528–35, 2001.

[18] Rossi FM, Kringstein AM, Spicher A, Guicherit OM, Blau HM. Transcriptional control: rheostat converted to on/off switch. *Mol Cell*, 6(3):723–8, 2000.

[19] Alper H, Fischer C, Nevoigt E, Stephanopoulos G. Tuning genetic control through promoter engineering. *Proc Natl Acad Sci U S A*, 102(36):12678–83, 2005.

[20] Beinoraviciūte-Kellner R, Lipps G, Krauss G. In vitro selection of dna binding sites for abf1 protein from saccharomyces cerevisiae. *FEBS Lett.*, 579(20):4535–40, 2005.

[21] Murphy KF, Balázsi G, Collins JJ. Combinatorial promoter design for engineering noisy gene expression. *Proc Natl Acad Sci U S A*, 104(31):12726–31, 2007.

[22] Wang HH, Isaacs FJ, Carr PA, Sun ZZ, Xu G, Forest CR, Church GM. Programming cells by multiplex genome engineering and accelerated evolution. *Nature*, 460(7257):894–8, 2009.

[23] Wissmann A, Meier I, Hillen W. Saturation mutagenesis of the tn10-encoded tet operator o1. identification of base-pairs involved in tet repressor recognition. *J Mol Biol*, 202(3):397–406, 1988.

[24] Adair GS. The hemoglobin system. vi. the oxygen dissociation curve of hemoglobin. *J Biol Chem*, 63:529–545, 1925.

[25] Giniger E, Ptashne M. Cooperative dna binding of the yeast transcriptional activator gal4. *Proc Natl Acad Sci U S A*, 85(2):382–6, 1988.

[26] Vashee S, Kodadek T. The activation domain of gal4 protein mediates cooperative promoter binding with general transcription factors in vivo. *Proc Natl Acad Sci U S A*, 92(23):10683–7, 1995.

[27] Vashee S, Xu H, Johnston SA, Kodadek T. How do "zn2 cys6" proteins distinguish between similar upstream activation sites? comparison of the dna-binding specificity of the gal4 protein in vitro and in vivo. *J Biol Chem*, 268(33):24699–706, 1993.

[28] Carey M, Lin YS, Green MR, Ptashne M. A mechanism for synergistic activation of a mammalian gene by gal4 derivatives. *Nature*, 345(6273):361–4, 1990.

[29] Lin YS, Carey MF, Ptashne M, Green MR. Gal4 derivatives function alone and synergistically with mammalian activators in vitro. *Cell*, 54(5):659–64, 1988.

[30] http://www.ornl.gov/sci/techresources/human_genome/project/info.shtml.

[31] Gertz J, Siggia ED, Cohen BA. Analysis of combinatorial cis-regulation in synthetic and genomic promoters. *Nature*, 457(7226):215–8, 2009.

[32] Bhattacharyya RP, Reményi A, Good MC, Bashor CJ, Falick AM, Lim WA. The ste5 scaffold allosterically modulates signaling output of the yeast mating pathway. *Science*, 311(5762):822–6, 2006.

[33] Baur M, Esch RK, Errede B. Cooperative binding interactions required for function of the ty1 sterile responsive element. *Mol Cell Biol*, 17(8):4330–7, 1997.

[34] Halme A, Bumgarner S, Styles C, Fink GR. Genetic and epigenetic regulation of the flo gene family generates cell-surface variation in yeast. *Cell*, 116(3):405–15, 2004.

[35] Polach KJ, Widom J. A model for the cooperative binding of eukaryotic regulatory proteins to nucleosomal target sites. *J Mol Biol*, 258(5):800–12, 1996.

[36] Segal E, Widom J. From dna sequence to transcriptional behaviour: a quantitative approach. *Nat Rev Genet*, 10(7):443–56, 2009.

[37] Floer M, Bryant GO, Ptashne M. Hsp90/70 chaperones are required for rapid nucleosome removal upon induction of the gal genes of yeast. *Proc Natl Acad Sci U S A*, 105(8):2975–80, 2008.

[38] Ashe HL, Briscoe J. The interpretation of morphogen gradients. *Development*, 133(3):385–94, 2006.

[39] Field Y, Kaplan N, Fondufe-Mittendorf Y, Moore IK, Sharon E, Lubling Y, Widom J, Segal E. Distinct modes of regulation by chromatin encoded through nucleosome positioning signals. *PLoS Comput Biol*, 4(11):e1000216, 2008.

[40] Badis G, Berger MF, Philippakis AA, Talukder S, Gehrke AR, Jaeger SA, Chan ET, Metzler G, Vedenko A, Chen X, Kuznetsov H, Wang CF, Coburn D, Newburger DE, Morris Q, Hughes TR, Bulyk ML. Diversity and complexity in dna recognition by transcription factors. *Science*, 324(5935):1720–3, 2009.

[41] Chapman SA, Asthagiri AR. Quantitative effect of scaffold abundance on signal propagation. *Mol Syst Biol*, 5:313, 2009.

[42] Lelandais G, Saint-Georges Y, Geneix C, Al-Shikhley L, Dujardin G, Jacq C. Spatio-temporal dynamics of yeast mitochondrial biogenesis: transcriptional and post-transcriptional mrna oscillatory modules. *PLoS Comput Biol*, 5(6):e1000409, 2009.

[43] Novák B, Tyson JJ. Design principles of biochemical oscillators. *Nat Rev Mol Cell Biol*, 9(12):981–91, 2008.

[44] Gerold Baier, Sven Sahle. Spatio-temporal patterns with hyperchaotic dynamics in diffusively coupled biochemical oscillators. *Discrete Dynamics in Nature and Society*, 1(2):161–167, 1997.

[45] Kelemen JZ, Ratna P, Scherrer S, Becskei A. Spatial epigenetic control of mono- and bistable gene expression. *PLoS Biol*, 8(3)::e1000332, 2010.

[46] Becskei A, Serrano L. Engineering stability in gene networks by autoregulation. *Nature*, 405(6786):590–3, 2000.

[47] Gardner TS, Cantor CR, Collins JJ. Construction of a genetic toggle switch in escherichia coli. *Nature*, 403(6767):339–42, 2000.

[48] Ferrell JE Jr. Self-perpetuating states in signal transduction: positive feedback, double-negative feedback and bistability. *Curr Opin Cell Biol*, 14(2):140–8, 2002.

[49] Ozbudak EM, Thattai M, Lim HN, Shraiman BI, Van Oudenaarden A. Multistability in the lactose utilization network of escherichia coli. *Nature*, 427(6976):737–40, 2004.

[50] Becskei A. Linearization through distortion: a new facet of negative feedback in signalling. *Mol Syst Biol*, 5:255, 2009.

[51] Nevozhay D, Adams RM, Murphy KF, Josic K, Balázsi G. Negative autoregulation linearizes the dose-response and suppresses the heterogeneity of gene expression. *Proc Natl Acad Sci U S A*, 106(13):5123–8, 2009.

[52] Schumacher MA, Miller MC, Grkovic S, Brown MH, Skurray RA, Brennan RG. Structural basis for cooperative dna binding by two dimers of the multidrug-binding protein qacr. *EMBO J*, 21(5):1210–8, 2002.

[53] Eriksson PR, Mendiratta G, McLaughlin NB, Wolfsberg TG, Mariño-Ramírez L, Pompa TA, Jainerin M, Landsman D, Shen CH, Clark DJ. Global regulation by the yeast spt10 protein is mediated through chromatin structure and the histone upstream activating sequence elements. *Mol Cell Biol*, 25(20):9127–37, 2005.

[54] Mendiratta G, Eriksson PR, Clark DJ. Cooperative binding of the yeast spt10p activator to the histone upstream activating sequences is mediated through an n-terminal dimerization domain. *Nucleic Acids Res*, 35(3):812–21, 2007.

[55] http://www.yeastgenome.org/.

[56] D. B. Nikolov and S. K. Burley. Rna polymerase ii transcription initiation: A structural view. *Proc Natl Acad Sci U S A*, 94(1):1522, 1997.

[57] Merry S, Britton HG. The mechanism of rabbit muscle phosphofructokinase at ph8. *Biochem J*, 226(1):13–28, 1985.

[58] Ricard J, Noat G, Got C, Borel M. The theory of alternative substrates in enzyme kinetics and its application to yeast hexokinase. *Eur J Biochem*, 31(1):14–24, 1972.

[59] Smith J, Stein V. Sporcalc: A development of a database analysis that provides putative metabolic enzyme reactions for ligand-based drug design. *Comput Biol Chem*, 33(2):149–59, 2009.

[60] Cronin CN, Tipton KF. Kinetic studies on the reaction catalysed by phosphofructokinase from trypanosoma brucei. *Biochem J*, 245(1):13–8, 1987.

[61] Arisaka F, Van Holde KE. Allosteric properties and the association equilibria of hemocyanin from callianassa californiensis. *J Mol Biol*, 15;134(1):41–73, 1979.

[62] Zhou T, Caflisch A. High-throughput virtual screening using quantum mechanical probes: discovery of selective kinase inhibitors. *ChemMedChem*, 5(7):1007–14, 2010.

[63] Eisenthal R, Cornish-Bowden A. Prospects for antiparasitic drugs. the case of trypanosoma brucei, the causative agent of african sleeping sickness. *J Biol Chem*, 273(10):5500–5, 1998.

[64] Giordani R, Buc J, Cornish-Bowden A, Cárdenas ML. Kinetics of membrane-bound nitrate reductase a from escherichia coli with analogues of physiological electron donors–different reaction sites for menadiol and duroquinol. *Eur J Biochem*, 250(2):567–77, 1997.

[65] Bagheri HC, Wagner GP. Evolution of dominance in metabolic pathways. *Genetics*, 168(3):1713–35, 2004.

[66] Ekonomiuk D, Su XC, Ozawa K, Bodenreider C, Lim SP, Yin Z, Keller TH, Beer D, Patel V, Otting G, Caflisch A, Huang D. Discovery of a non-peptidic inhibitor of west nile virus ns3 protease by high-throughput docking. *PLoS Negl Trop Dis*, 3(1), 2009.

[67] Mokbel K. The evolving role of aromatase inhibitors in breast cancer. *Int J Clin Oncol*, 7(5):279–83, 2002.

[68] Kolb P, Huang D, Dey F, Caflisch A. Discovery of kinase inhibitors by high-throughput docking and scoring based on a transferable linear interaction energy model. *J Med Chem*, 51(5):1179–88, 2008.

[69] Hanahan D, Weinberg RA. The hallmarks of cancer. *Cell*, 100(1):57–70, 2000.

[70] J.D. Murray. *Mathematical biology. I. An introduction.* Springer, Springer-Verlag Berlin Heidelberg, 2002.

[71] Christopher P. Fall. *Computational cell biology.* Springer, Springer-Verlag New York, Inc., 2002.

[72] A Cornish Bowden. *Fundamentals of Enzyme Kinetics.* Portland Press Ltd, 2004.

[73] Wienken CJ, Baaske P, Rothbauer U, Braun D, Duhr S. Protein-binding assays in biological liquids using microscale thermophoresis. *Nat Commun*, 1(7):100, 2010.

[74] Bancaud A, Huet S, Daigle N, Mozziconacci J, Beaudouin J, Ellenberg J. Molecular crowding affects diffusion and binding of nuclear proteins in heterochromatin and reveals the fractal organization of chromatin. *EMBO J*, 28(24):3785–98, 2009.

[75] Magno A, Amedeo Caflisch A, Pellarin R. Crowding effects on amyloid aggregation kinetics. *J Phys Chem Lett*, 1 (20):30273032, 2010.

[76] Savageau MA. Michaelis-menten mechanism reconsidered: implications of fractal kinetics. *J Theor Biol*, 176(1):115–24, 1995.

[77] Becskei A, Kaufmann BB, van Oudenaarden A. Contributions of low molecule number and chromosomal positioning to stochastic gene expression. *Nat Genet*, 37(9):937–44, 2005.

[78] McAdams HH, Arkin A. It's a noisy business! genetic regulation at the nanomolar scale. *Trends Genet*, 15(2):65–9, 1999.

[79] Wilkinson DJ. Stochastic modelling for quantitative description of heterogeneous biological systems. *Nat Rev Genet*, 10(2):122–33, 2009.

[80] Vershon AK, Johnson AD. A short, disordered protein region mediates interactions between the homeodomain of the yeast alpha 2 protein and the mcm1 protein. *Cell*, 72(1):105–12, 1993.

[81] Bintu L, Buchler NE, Garcia HG, Gerland U, Hwa T, Kondev J, Phillips R. Transcriptional regulation by the numbers: models. *Curr Opin Genet Dev*, 15(2):116–24, 2005.

[82] Bintu L, Buchler NE, Garcia HG, Gerland U, Hwa T, Kondev J, Kuhlman T, Phillips R. Transcriptional regulation by the numbers: applications. *Curr Opin Genet Dev*, 15(2):125–35, 2005.

[83] Yamada T, Bork P. Evolution of biomolecular networks lessons from metabolic and protein interactions. *Nat Rev Mol Cell Biol*, 10(11):791–803, 2009.

[84] Briggs GE, Haldane JB. A note on the kinetics of enzyme action. *Biochem J*, 19(2):338–9, 1925.

[85] Hill AV. The possible effect of the aggregation of the molecules of haemoglobin on its dissociation curves. *J Physiol*, 40:4–7, 1910.

[86] Inc. Clontech Laboratories. Yeast protocols handbook. pages 28–29, 2009.

[87] Daniel M. Becker and Victoria Lundblad . Current protocols in molecular biology. pages 13.7.1–13.7.10, 1993.

[88] Douglas A. Treco and Victoria Lundblad. Current protocols in molecular biology. pages 13.1.1–13.1.7, 1993.

[89] Michael Finney, Paul E. Nisson and Ayoub Rashtchian. Current protocols in molecular biology. pages 15.4.1–15.4.11, 2001.

[90] Urlinger S, Baron U, Thellmann M, Hasan MT, Bujard H, Hillen W. Exploring the sequence space for tetracycline-dependent transcriptional activators: novel mutations yield expanded range and sensitivity. *Proc Natl Acad Sci U S A*, 97(14):7963–8, 2000.

[91] Foster TJ, Davis MA, Roberts DE, Takeshita K, Kleckner N. Genetic organization of transposon tn10. *Cell*, 23(1):201–13, 1981.

[92] Beck CF, Mutzel R, Barbé J, Müller W. A multifunctional gene (tetr) controls tn10-encoded tetracycline resistance. *J Bacteriol*, 150(2):633–42, 1982.

[93] Bertrand KP, Postle K, Wray LV Jr, Reznikoff WS. Overlapping divergent promoters control expression of tn10 tetracycline resistance. *Gene*, 23(2):149–56, 1983.

[94] Hillen W, Schollmeier K, Gatz C. Control of expression of the tn10-encoded tetracycline resistance operon. ii. interaction of rna polymerase and tet repressor with the tet operon regulatory region. *J Mol Biol*, 172(2):185–201, 1984.

[95] Louvion JF, Havaux-Copf B, Picard D. Fusion of gal4-vp16 to a steroid-binding domain provides a tool for gratuitous induction of galactose-responsive genes in yeast. *Gene*, 131(1):129–34, 1993.

[96] Ratna P, Scherrer S, Fleischli C, Becskei A. Synergy of repression and silencing gradients along the chromosome. *J Mol Biol*, 387(4):826–39, 2009.

[97] McBride HJ, Yu Y, Stillman DJ. Distinct regions of the swi5 and ace2 transcription factors are required for specific gene activation. *J Biol Chem*, 274(30):21029–36, 1999.

[98] Krebs JE, Fry CJ, Samuels ML, Peterson CL. Global role for chromatin remodeling enzymes in mitotic gene expression. *Cell*, 102(5):587–98, 2000.

[99] S Lawrence Zipursky Paul Matsudaira David Baltimore Harvey Lodish, Arnold Berk and James Darnell. *Molecular Cell Biology*. W. H. Freeman and Company, 2000.

[100] Kovacech B, Nasmyth K, Schuster T. Egt2 gene transcription is induced predominantly by swi5 in early g1. *Mol Cell Biol*, 16(7):3264–74, 1996.

[101] Cheng JX, Floer M, Ononaji P, Bryant G, Ptashne M. Responses of four yeast genes to changes in the transcriptional machinery are determined by their promoters. *Curr Biol*, 12(21):1828–32, 2002.

[102] Dhasarathy A, Kladde MP. Promoter occupancy is a major determinant of chromatin remodeling enzyme requirements. *Mol Cell Biol*, 25(7):2698–707, 2005.

[103] Tigges M, Marquez-Lago TT, Stelling J, Fussenegger M. A tunable synthetic mammalian oscillato.r. *Nature*, 457(7227):309–12, 2009.

[104] Hasty J, McMillen D, Collins JJ. Engineered gene circuits. *Nature*, 420(6912):224–30, 2002.

[105] McAdams HH, Arkin A. Stochastic mechanisms in gene expression. *Proc Natl Acad Sci U S A*, 94(3):814–9, 1997.

[106] Hartwell LH, Hopfield JJ, Leibler S, Murray AW. From molecular to modular cell biology. *Nature*, 402(6761 Suppl):C47–52, 1999.

[107] Lauffenburger DA. Cell signaling pathways as control modules: complexity for simplicity? *Proc Natl Acad Sci U S A*, 97(10):5031–3, 2000.

[108] Orth P, Schnappinger D, Hillen W, Saenger W, Hinrichs W. Structural basis of gene regulation by the tetracycline inducible tet repressor-operator system. *Nat Struct Biol*, 7(3):215–9, 2000.

[109] Melcher K, Xu HE. Gal80-gal80 interaction on adjacent gal4p binding sites is required for complete gal gene repression. *EMBO J*, 20(4):841–51, 2001.

[110] Johnston M. A model fungal gene regulatory mechanism: the gal genes of saccharomyces cerevisiae. *Microbiol Rev*, 51(4):458–76, 1987.

[111] Reece RJ, Ptashne M. Determinants of binding-site specificity among yeast c6 zinc cluster proteins. *Science*, 261(5123):909–11, 1993.

[112] Hillen W, Berens C. Mechanisms underlying expression of tn10 encoded tetracycline resistance. *Annu Rev Microbiol*, 48:345–69, 1994.

[113] Krueger C, Berens C, Schmidt A, Schnappinger D, Hillen W. Single-chain tet transregulators. *Nucleic Acids Res*, 31(12):3050–6, 2003.

[114] Acar M, Becskei A, van Oudenaarden A. Enhancement of cellular memory by reducing stochastic transitions. *Nature*, 435(7039):228–32, 2005.

[115] Daniel T. Gillespie. A general method for numerically simulating the stochastic time evolution of coupled chemical reactions. *Journal of Computational Physics*, 22(4):403–434, 1976.

[116] To TL, Maheshri N. Noise can induce bimodality in positive transcriptional feedback loops without bistability. *Science*, 327(5969):1142–5, 2010.

[117] Edgar Serfling, Maria Jasin and Walter Schaffner. Enhancers and eukaryotic gene transcription. *Trends in Genetics*, 1:224–230, 1985.

[118] Raab JR, Kamakaka RT. Insulators and promoters: closer than we think. *Nat Rev Genet*, 11(6):439–46, 2010.

i want morebooks!

Buy your books fast and straightforward online - at one of world's fastest growing online book stores! Environmentally sound due to Print-on-Demand technologies.

Buy your books online at
www.get-morebooks.com

Kaufen Sie Ihre Bücher schnell und unkompliziert online – auf einer der am schnellsten wachsenden Buchhandelsplattformen weltweit! Dank Print-On-Demand umwelt- und ressourcenschonend produziert.

Bücher schneller online kaufen
www.morebooks.de

VDM Verlagsservicegesellschaft mbH
Heinrich-Böcking-Str. 6-8   Telefon: +49 681 3720 174   info@vdm-vsg.de
D - 66121 Saarbrücken   Telefax: +49 681 3720 1749   www.vdm-vsg.de

Printed by Books on Demand GmbH, Norderstedt / Germany